GREAT LAKES WETLAND WALKS

A WILDFLOWER FIELD GUIDE

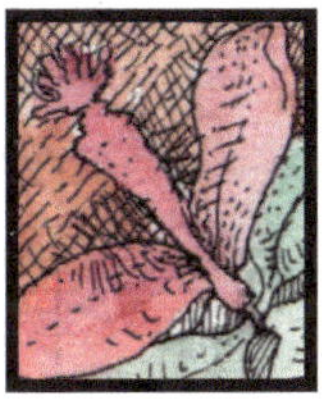

PEG COMFORT

FOREWORD BY JERRY DENNIS
PHOTOGRAPHS BY JAMES DAKE

PUBLISHED BY
GRASS RIVER
NATURAL AREA
BELLAIRE, MICHIGAN
2019

GRASS RIVER
NATURAL AREA
50
1969 – 2019

Grass River Natural Area, Inc. Special Publication No. 2

Great Lakes Wetland Walks:
A Wildflower Field Guide

ISBN-13: 978-1-7341277-0-6
ISBN-10: 1-7341277-0-8
Library of Congress Control Number: 2019951959

2nd printing, 2022

The printing of this book was made possible by:
Grass River Natural Area, Inc.
Grand Traverse Regional Community Foundation
The Dole Family Foundation

Published by:
Grass River Natural Area, Inc.
PO BOX 231 - Bellaire, MI 49615
www.grassriver.org
231-533-8314

Cover Design by Glenn Wolff
Photographs by James Dake
Plant Diagrams by Heather Shaw

Printed in the U.S.A.

TABLE OF CONTENTS

DEDICATION

To Sue Schuler - my dear friend, an exceptional naturalist, passionate plant lover, skilled teacher and superb writer.

Thank you for sharing many wonderful wetland walks, spending hours helping to craft this beginner's wildflower guide and spreading your love for our beautiful Earth!

I can't express the depth of my gratitude to you!

With love - Peg

ACKNOWLEDGMENTS

David G. Peterson, Jr., my "big brother," gave me my first flower guide to Rocky Mountain wildflowers. Little did I know this book would lead to a life filled with a passion for plants.

Ruth Cook, Margaret Campbell and Caroline Hoadley knew the value of wild places and possessed the courage to help create the Grass River Natural Area in Antrim County, Michigan. Thank you for sharing this wetland wonder with me and inviting me into the local conservation community.

Shirley Mills shared early explorations of the Grass River Natural Area-before there were trails and thousands of visitors. Together, we adventured into the mosaic of habitats in these wonderful wetlands. As early Board members, we helped to protect them for future generations to enjoy, led field trips and laid the foundation for the educational program.

Helen Mary Powell, Sue Schuler and Sally Malone loved summer adventures in the Elk River Chain of Lakes Watershed. The beauty we encountered together was matched by your spirits as "Woods Women."

And most of all - Tom Comfort, the love of my life. Without your encouragement, patience, understanding and support, I wouldn't have gotten this book to print.

Much love to you all, Peg

FOREWORD

Whenever I start to abandon hope for the planet I hightail it to a few hidden places where I can lose myself—for, of course, losing ourselves is the surest way to find ourselves. Many of those places are within the swamps, bogs, fens, and marshes that most people have always been determined to avoid. They avoid them because they assume such places are inhospitable, uncongenial, even treacherous.

Those who do go, however, are wise to learn as much as they can about them. Many of us discovered as kids that when we learn the names of things our eyes open wider and we begin to really see the world around us. The diversity of life in a wetlands—and all the replenishment and proliferation that occurs there—is why they are among the few habitats that we've consistently protected for the past half century. These places that serve so beneficially as nurseries for plants and animals are nurseries for hope, as well.

Among my own hopes is that anyone with an adventurous spirit and a hunger to learn will carry Peg Comfort's book everywhere they go. It's a field guide, after all, and should be taken afield. I know how my own copy will look a year from now: water-wrinkled and dog-eared, with mud on the cover, a few mosquitoes pressed between the pages, and many scribbled notes inside. That's as it should be. Another thing we learned as kids is that immersing ourselves in the world is a deliciously messy business.

Jerry Dennis

INTRODUCTION

Dear Reader,

Wetlands are wild places, filled with tangled trees and shrubs, deep muck and beautiful native plants. They're all around us but often overlooked, because they aren't always accessible. Now, many preserves throughout the Great Lakes Watershed have miles of boardwalks, making them easier to explore.

You'll love exploring wetlands! Whether you go to your favorite haunts or to new places, you know a new adventure will present itself. How do I know that? Each new flower is like a flag that stops you in your tracks. Zoom in like a camera to focus on the details—maybe on your hands and knees, maybe with a magnifying glass. By interacting with the plant in this way, you create a special relationship, enabling you to remember the plant for years to come. You may get so absorbed you'll completely lose track of time. The diversity, intricacy and beauty of nature will simply amaze you!

I am excited to share our **<u>GREAT LAKES WETLAND WALKS</u>** with you!

The Plant Guide contains pictures of common wetland flowers you're likely to encounter in the Great Lakes Watershed. I've included a process for getting "up close and personal" with these flowers. You're not just looking at and identifying a flower. I am encouraging you to look more closely at the whole plant—its flowers, leaves and stem. This way each plant can become a friend to treasure for a lifetime. This approach will make wetland walks memorable.

I've also included the natural history of wetlands. Their history helps you understand which plants grow together and how plant communities are part of a larger, wetland ecosystem. Wetlands are important but they are becoming scarce. Already more than 75% of the wetlands in the Great Lakes Watershed have been destroyed. Our future is connected to the health of wetlands, so we need to each do what we can to protect them.

Great Lakes Wetland Walks is:

- Easy to Use - Wetland flowers are arranged by seasons: spring, early summer and late summer. Orchids are discussed with additional detail in Chapter 5. I've included a step-by-step process for identifying common flowers with a limited number of scientific words.

- Fun in the Field - Get up close and personal. Look at it. Touch it. Smell it. Look for patterns and features you've never noticed before.

- Interactive - Use the plant guide as your companion in the field. Make notes and sketches. They will help you remember plants that you meet on your walks.

- Focused - The pictures are limited to the most common wetland plants found in the most common wetland plant communities.

- A Resource Guide - Learn about how plants changed the world, plant lists, glossary, photo credits and index.

Have fun searching for the wildflowers and other amazing wetlands plants in our Great Lakes Watershed!

Peg Comfort

P.S. Please share your adventures, discoveries and pictures by visiting: www.grassriver.org

Know the Law!

Many wetland plants are protected by state and federal laws.

1) Don't pick'em!

2) Don't dig'em!

Beware of toxic plants!

Don't touch'em! They may cause skin rashes and severe allergic reactions.

Poison ivy
(Leaves of three; let them be!)

Poison sumac
(Berries white; take flight!)

Stinging nettle
(Stinging hairs; beware!)

Leafy spurge
(Stem & leaves that ooze; let them be!)

Don't eat'em! Plants or berries may be poisonous.

Showy lady's slipper

Deadly nightshade

Pack your bag!

Before you head out on your walk, please tell someone where you are going and when you'll be back. Better yet, take a buddy! Be sure you have everything you need:

Backpack
Layers: tee shirt, long-sleeved shirt and jacket
Fluids: water, insect repellent and sunscreen
Tools: pencil, **Great Lakes Wetland Walks**,
magnifier, binoculars and camera
Snacks: nothing like good snacks to fuel your hike

PLANT
GUIDE

CHAPTER ONE
APPRECIATING WETLANDS

Map of the Great Lakes Watersheds

What is a wetland?

Great Lakes Watershed - There are more than 36 different kinds of wetlands in the Great Lakes Watershed, with the most common wetlands being cedar swamp forest, shrub thicket, northern wet meadow and emergent marsh.

Soggy, squishy & swampy places - Wetlands are important because they provide us with clean water and clean air. Depending on the conditions and the location, each wetland has its own plant community providing critical habitat for butterflies, insects, reptiles, amphibians, birds, and mammals.

What are the characteristics of wetlands?

Water - Wetlands need water and lots of it! The amount of water and how it moves determines the type of wetland.

> *Marshes and wet meadows* develop in shallow water along the edges of lakes and rivers where currents and waves keep water moving.
>
> *Bogs and fens* form where water levels change seasonally, sometimes isolated and sometimes in pockets located within larger swamp forests.
>
> *Shrub thickets and cedar swamp forests* grow in standing water between the lake and uplands.

Soil - Mucky, soggy soil describes wetlands. But not all wetland soils are alike. Some hold more water than others. Some are more acidic or more basic. The type of soil and amount of water in the soil affect the plants that grow. For example, bogs are very acidic, lacking nutrients, so plants are short and small. Fens are more basic, so plants are taller and larger.

Plants - Wetland plants have adapted to living in standing or flowing water year-round.

Plants have also adapted to different light conditions. For example, wet meadows and marshes have sunlight all day long, but cedar swamps are shaded by trees. Plants in low light are smaller with larger leaves or many small leaves, while those in full sun grow taller and stronger.

How did wetlands get here?

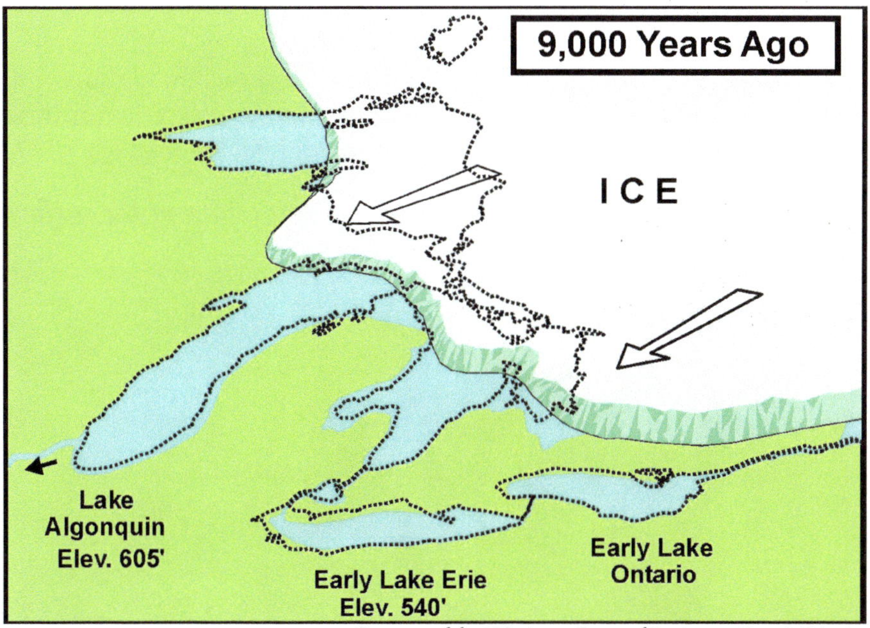

Courtesy of the U.S. Army Corps of Engineers, Detroit District

Glaciers - For thousands of years, mile-high sheets of ice called glaciers covered most of the world north of the 45th parallel in North America, Europe and Asia. The glaciers grew bigger as they advanced, and became smaller as they melted and retreated many times. They formed wetlands in the Great Lakes Watershed.

Surface and Ground Water - Each time the glaciers advanced, the ice gouged the bedrock of the continent and retreated, leaving behind freshwater and glacial till (cobblestones, ground up pebbles, sand and clay). This freshwater formed the surface water of the lakes and streams we find in the Great Lakes Watershed.

Much more freshwater was left in thick layers of glacial till. This groundwater provides us with drinking water and recharges surface water. Our soils are determined by the make-up of these glacial till deposits.

Wetlands - Old lakeshores of ancient glacial lakes provided mucky, wet soil and wet places that were colonized by plants adapted to those conditions. Over time, these plants formed unique plant communities which persist along our present day lakes, rivers and streams. These wetland plants are so well adapted to the northern climate and soils that they can't live anywhere else.

The glaciers gave many gifts to the Great Lakes region: tens of thousands of inland lakes and rivers, as well as streams and creeks. Each of these thousands of water bodies have wetlands along their shores.

Altogether, the Great Lakes Watershed contains 20% of the surface freshwater on Earth.

How do wetlands change over time?

A Mosaic - Different plant communities, each one distinct yet connected to those near it, make up wetlands. Instead of definite boundaries, edges overlap and one community blends into the next.

Succession - These communities are constantly changing in a process called succession, in which one stage of plants is succeeded by another. As you walk through wetlands, you'll see these different stages.

For example, when marsh plants grow along the edges of a lake, they die and accumulate on the bottom. This slows down the waves and currents and makes the lake shallower. Over time, the edge of the lake becomes soggy ground where sedges begin to take hold. In turn, they die and build up, making the ground firmer, so shrubs can grow. Flash forward a few thousand years, and trees start growing. Before you know it, the lake has become a swamp forest.

Wetland Plant Communities

Habitats - There are more than 36 different wetland habitats in the Great Lakes Region. This Plant Guide introduces you to four of the most common: Cedar Swamp Forest, Northern Shrub Thicket, Northern Wet Meadow and Emergent Marsh.

Trees and Shrubs - Trees and shrubs help to create two wetland plant communities: Cedar Swamp Forest and Northern Shrub Thicket. Trees and shrubs can form quite a tangle. Sometimes they're so thick and the ground is so wet, you can't even walk through them.

No Trees and Shrubs - Thank goodness for boardwalks! Boardwalks guide you through the jumble of blown down trees and squishy ground. When the boardwalk leads you into more sunlight, you'll notice grass-like plants joining or replacing trees and shrubs. Because of their flowing water, these plant communities are most easily seen by boat. They're called Northern Wet Meadow and Emergent Marsh.

How are wetland plant communities different?

Park your car where it is high and dry. Then start walking downhill. You may walk through sugar maple or white pine forests, open meadows or scrubby areas. As the ground gets wetter, the trees change from the upland forest (beech, maple, and hemlock) to the wetland forest (cedar, balsam fir, black spruce, and tamarack). Watch for other familiar trees and the plants beneath them signaling the change from dry to wet: sugar maple to red maple and white birch to yellow birch. Before you know it, you'll be standing at the water's edge in the sunshine.

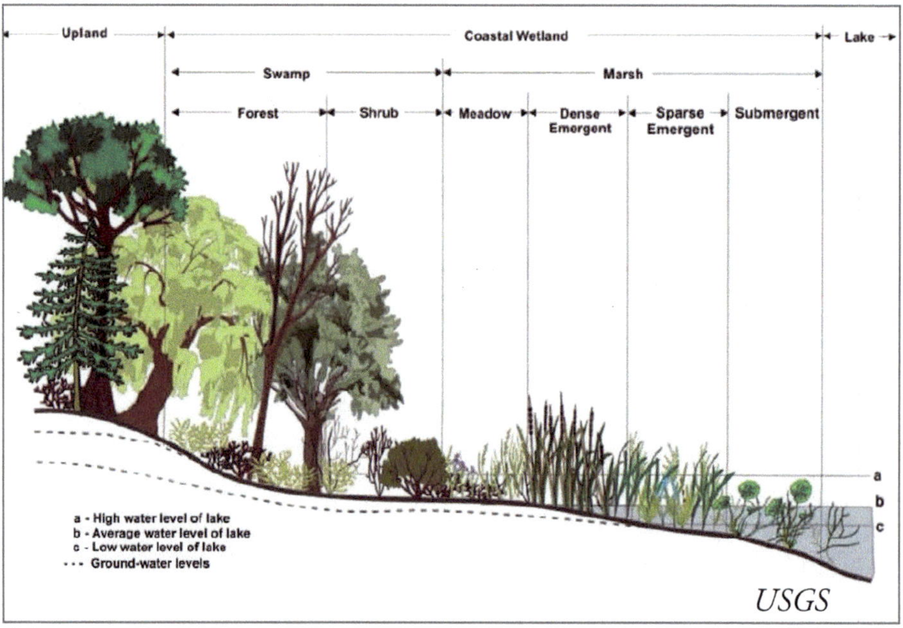

GREAT LAKES WETLANDS - Northern Basin

MICHIGAN

Lower Peninsula

Grass River Natural Area (Bellaire, MI)
Cheboygan State Park (Cheboygan, MI)
Jordan River State Forest (East Jordan, MI)
Skegemog Wildlife Area (Rapid City, MI)
Wilderness State Park (Carp Lake, MI)

Upper Peninsula

Escanaba River State Forest (Carney, MI)
Fumee Lake Natural Area (Iron Mountain, MI)
Pat Grogan Shelldrake Orchid Bog (Paradise, MI)
Pictured Rocks National Lakeshore (Munising, MI)
Porcupine Mountains Wilderness State Park (Ontonagon, MI)

MINNESOTA

Big Bog State Recreation Area (Wakish, MN)
Chippewa National Forest (Cass Lake, MN)
Itasca State Park (Park Rapids, MN)
Lake Bemidji State Park (Bemidji, MN)
Orr Bog Walk (Orr, MN)
Pennington Bog State Natural Area (Bemidji, MN)- MDNR permit required

WISCONSIN

Chequamegon-Nicolet National Forest (Park Falls, WI)
Flambeau River State Forest (Winter, WI)
Peninsula Park Beech Forest (Fish Creek, WI)
The Ridges Sanctuary (Bailey's Harbor, WI)
Whitefish Dunes State Park (Sturgeon Bay, WI)

NEW YORK

Beaver Lake Nature Center (Baldwinsville, NY)
Caleb Smith State Park Preserve (Smithtown, NY)
Constitution Marsh Audubon Center (Garrison, NY)
Five Rivers Environmental Education Center (Delmar, NY)
Huckleberry Swamp (Lyons, NY)
Jamestown Audubon Center (Jamestown, NY)
Newcomb Adirondack Visitor Center (Newcomb, NY)
Thousand Acre Swamp (Penfield, NY)

ONTARIO

Giant's Rib Discovery Centre (Hamilton, ON)
Guelph Lake Nature Centre (Guelph, ON)
Upper Canada Migratory Bird Sanctuary (Ingleside, ON)
Wye Marsh Wildlife Centre (Midland, ON)

CHAPTER TWO
LOOKING AT PLANTS

THE SIMPLE APPROACH

There are more than 2,000 wetland plants in the Great Lakes Watershed, BUT I've included only 100 of the most common wetland plants. This may seem like a lot, but this guide makes it easy to identify them.

So… you're walking in a wetland and come across a flowering plant that catches your eye. You wonder, "What is this?" Using the Simple Approach, you can find your answer in four steps.

Step One - What time of year are you walking in the wetland: late spring, early summer or late summer? Turn to the current season in the guide.

Step Two - What color is the flower you're observing? Find the appropriate color section within that season.

Step Three - Look through the pictures to find your plant.

Step Four - Read the description and make sure it matches the plant you are observing. You'll discover the plant's common name, as well as the scientific name and family.

Step Five - Use the Field Notes Page to record your observations. Make a sketch of the flower and leaves. You might want to record the date and location or information about other plants nearby.

Leatherleaf Family: Heath (Ericaceae)

Scientific Name: *Chamaedaphne calyculata* Height: 3' (0.9 m)

FLOWER: Many, tiny (¼-½"), regular, stalked, five lobed, fused into urn shape white flowers dangle from small-leaved, terminal branches. Flowers hang from leaf axils like little lanterns.

LEAF: Simple, alternate, short-stalked, oval, (1-2") leaves; upper surface brown-green, leathery; lower, pale with small rusty dots; edges smooth or with small round teeth. Leaves smallest on flowering branches.

STEM: Older stems gray, shedding to show red inner bark.

NOTE: Evergreen shrub, 3' high. Same family as bog rosemary and wintergreen. Compare urn-shape flowers and evergreen characteristics.

THE IN-DEPTH APPROACH

Learning the common and scientific name of a plant can be fun, but to really appreciate the uniqueness of a plant, you really have to spend some time observing more closely. The flowers, leaves and stem of each plant have special characteristics. Some are truly amazing!

There are many scientific terms used to describe flowers, leaves and stems. At first, they can be confusing and a little intimidating. I've included simple definitions and diagrams to make it easy for you to learn them. They provide a common vocabulary used in other field guides. If you take a wetland walk with a naturalist, then you'll hear these same terms, too. As you practice looking at plants with this In-Depth Approach, you'll become more familiar with not only the terms, but the patterns found in nature.

So… take some time to study a flowering plant in your yard. You'll be surprised by what you discover!

CAUTION! Make sure this plant is not toxic. Turn to Page 11-12

If the plant you are studying is a toxic plant, PLEASE DO NOT TOUCH!

LOOKING AT FLOWERS

Flowers are important to plants, because they're critical for reproduction. Flowers were created to attract specific pollinators (bees. butterflies and other insects). Each native plant has a specific native insect that is adapted to fertilize that specific flower AND to get the nutrition it needs for its own reproduction.

NOTE: Non-native plants and cultivars do not provide the specific nutrients needed for these native insects. Cultivars are native plants that have had some of their characteristics changed by humans to produce a specific flower color or characteristic.

Step One - The first thing you notice about a flower is its color. Colors can range from white, shades of pink to red, shades of yellow to orange and shades of blue to purple.

Observation: What color is the flower?

Step Two - The second characteristic you observe is the size of the flower. Sizes can range from very tiny to very large.

Observation: What size is the flower?

Flower Sizes

Very Tiny

Tiny (smaller than a dime)

Small (a dime)

Medium (a quarter),

Large (larger than a quarter)

Very Large

Step Three - Flowers can be organized into three (3) groups: regular, irregular and composite.

Observation: To which group does the flower you're observing belong?

Flower Groups

Regular Flowers

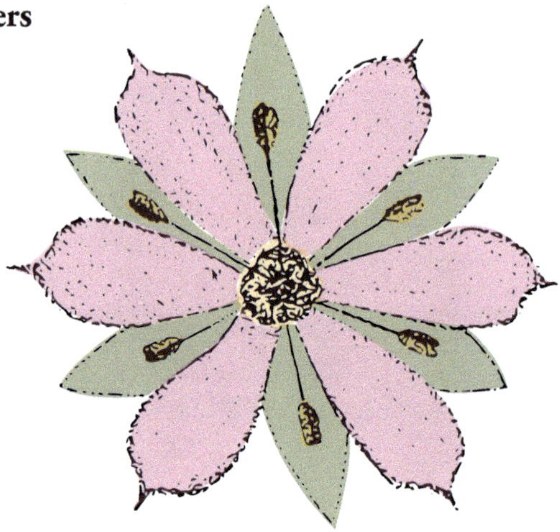

- Flowers such as forget-me-not, trillium, lily, and wild rose are regular flowers.
- Regular flowers have regular shapes: circle, urn, tube, or bell.
- Petals are arranged in a circle like spokes on a wheel (radial symmetry).
- All petals **are alike** in size, shape and color.
- Sometimes the petals are fused to form an urn, bell or tube.

Irregular Flowers

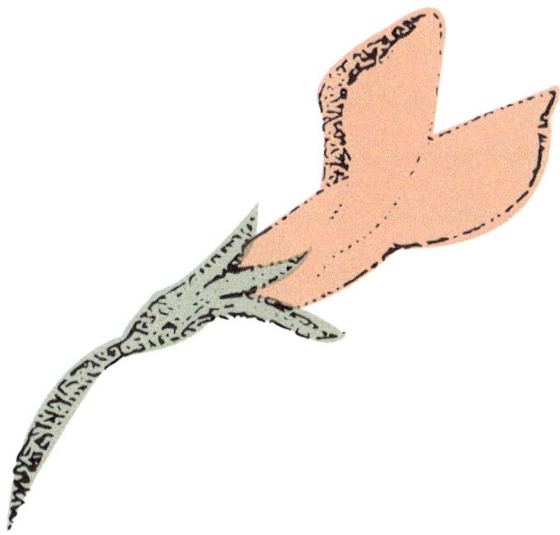

- Flowers such as orchids, mints, violets, snapdragons and peas have irregular flowers.
- Irregular flowers have an irregular shape.
- If the flower were cut in half vertically, both sides would be mirror images (bilateral symmetry).
- All flower parts **<u>are not alike</u>** in size, shape or color.

Composite Flowers

- More species of flowers belong to this group than either of the other two groups.
- Flowers such as daisies, sunflowers and goldenrods are composite flowers.
- The shape of the entire composite flower is a circle.
- A composite flower is really a cluster of tiny flowers bundled together to look like a single flower.
- Composite flowers have another distinguishing characteristic, an involucre. An involucre is a whorl of bracts that over-lap and form a cup-like vase.
- Composite flowers can be composed of one or both kinds of flowers—disk flowers and ray flowers.

Disk Flowers Only (thistle, Joe-Pye-weed, boneset) - tube-shaped with 5 pointed lobes, packed tightly together, fastened into a cup-like base formed by overlapping, small leaf-like structures called bracts

Ray Flowers Only (dandelion) - flat and narrow, ribbon-shaped, look like typical flower petals

Both Disk & Ray Flowers (aster)

Step Four - Flowers have four (4) basic shapes: circular, irregular, urn and bell.

Observation: Which flower shape best describes the flower you're observing?

Flower Shapes

Circular

Bell

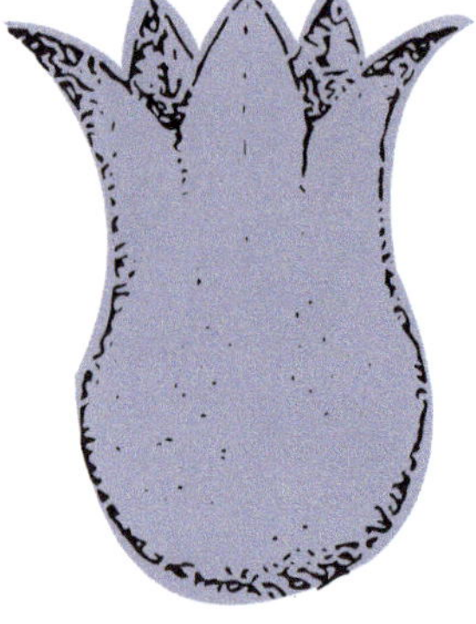

Urn

Irregular

Step Five - Sometimes flowers are arranged in clusters. There are three (3) common clusters shapes: spike-like, flat-top and round-top.

Observation: If the flowers in the plant you are observing are arranged in a cluster, which shape is the cluster?

<u>Flower Cluster Shapes</u>

Spike-like

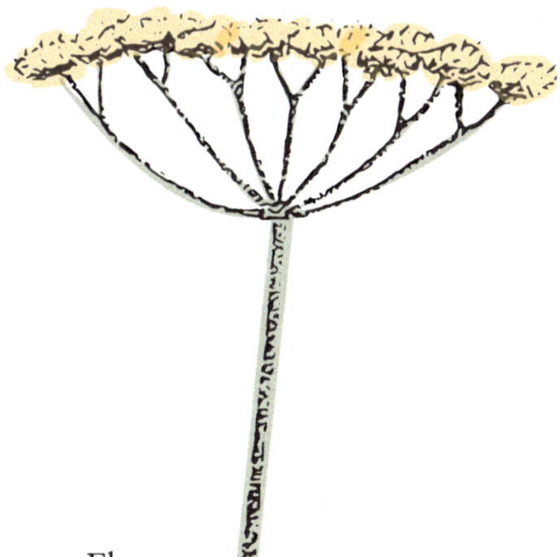

Flat-top

Round-top

Step Six - Flowers have two (2) basic parts: petals and sepals. Sometimes it's hard to tell them apart. The flower stalk attaches the flower to the stem.

Observation: How many petals and how many sepals does the flower you are observing have?

Urn, bell and tube-shaped flowers have fused petals that form lobes. For these flowers, count the outer edge of the flower (lobes) to determine the number of petals.

Flower Parts

PETALS- usually the colored part of the flower, may be fused at the base or separate.

SEPALS- the outermost parts of a flower, encloses the bud, and after the bud opens, they're found below the petals; not all flowers have petals. Sometimes the sepals are colored and mistaken for petals.

Step Seven - Flowers can reproduce because they have male and female reproductive parts as described below.

Observation: How many anthers does your flower have?

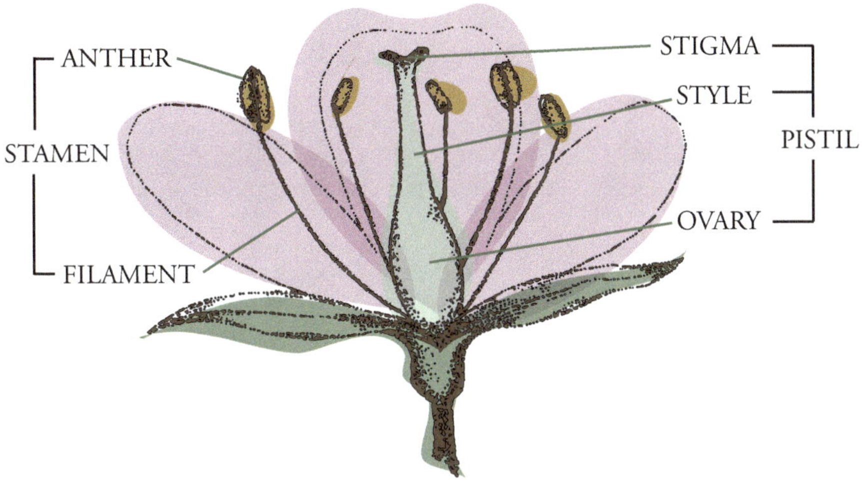

Flower Reproductive Parts

MALE PARTS

STAMEN - male reproductive organ usually surrounding the pistil, contains the anther and filament, and produces the pollen

ANTHER - enlarged part of the stamen holding the pollen

FILAMENT - the anther-bearing stalk of the stamen

POLLEN - the male spores produced by the anther

FEMALE PARTS

PISTIL- female reproductive organ located in the center of flower, contains the stigma, style and ovary which produces the seed(s)

OVARY- the swollen base of the pistil where seeds develop

STIGMA- the sticky tip of the pistil

STYLE- the neck of the pistil

LOOKING AT LEAVES

Leaves are important to plants because they capture the sun's light energy and change it to chemical energy the plant needs to grow and reproduce. This process is called photosynthesis.

Sometime leaf characteristics can help you identify a plant, even when it's not blooming.

Step One - Leaves can be organized into four (4) basic groups: simple, divided, compound-pinnate and compound-palmate.

Observation: To which leaf group does the plant you're observing belong?

Leaf Groups

Simple Leaf

- A single leaf attached to the plant stem
- Attached with or without a leaf stalk
- Attached individually to the plant stem

Divided Leaf

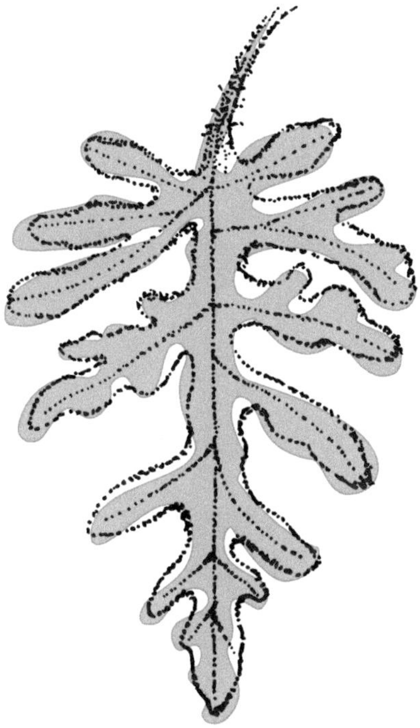

- A single leaf attached to the plant stem
- Attached with or without a leaf stalk
- May be deeply divided or cut to the central vein of the leaf or into lobes
- Maybe be divided or compound

Compound Leaf

- Has many separate divisions that look like small leaves, called leaflets
- Attached to the central leaf stem and NOT to the plant stem
- May be arranged in two (2) ways on the leaf stem: pinnate or palmate

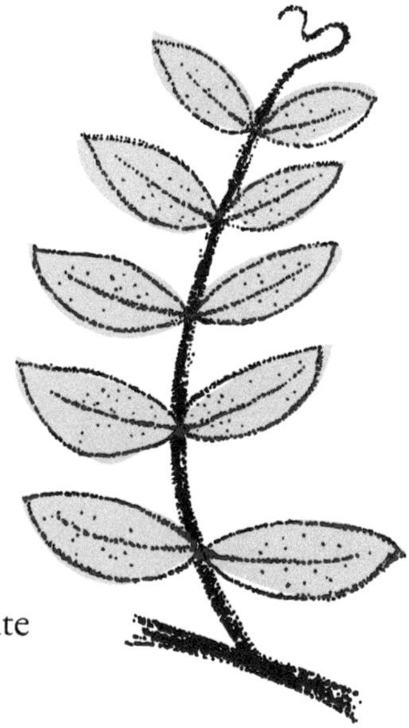

Pinnate

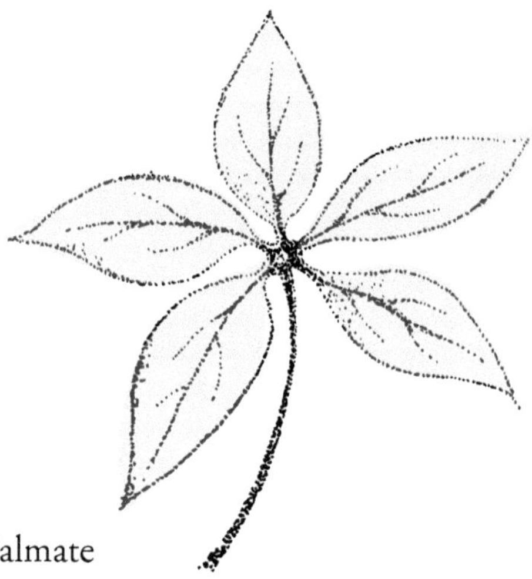

Palmate

Step Two - Leaves have five (5) basic shapes: narrow, lance shape, oval shape, heart shape, and round/kidney shape.

Observation: What is the basic shape of the leaf you are observing?

<u>Leaf Shapes</u>

Narrow

Lance shape

Oval shape

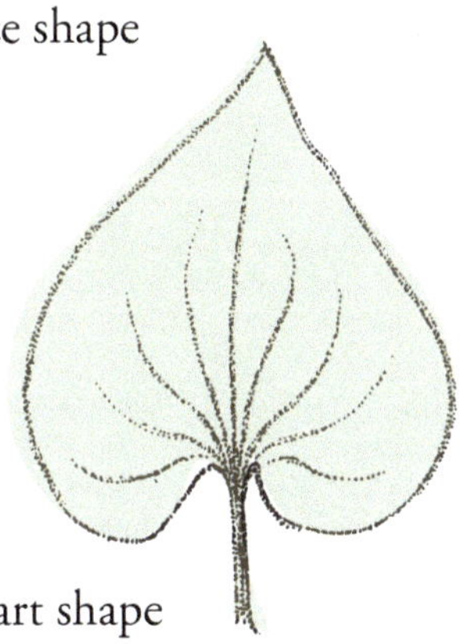

Heart shape

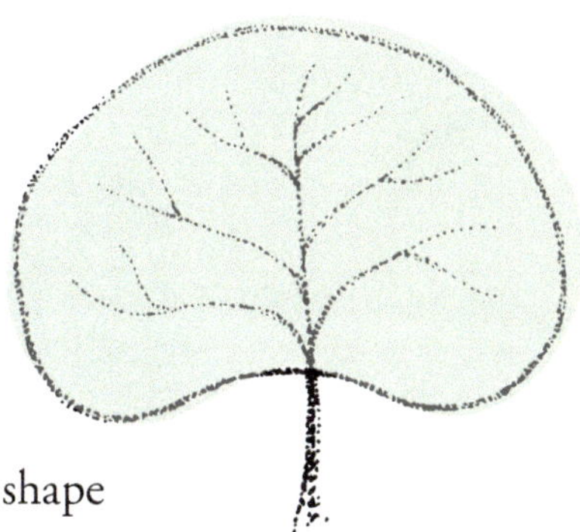

Round/kidney shape

Step Three - Leaves can be arranged in four (4) basic ways: alternate, opposite, whorled or basal rosette.

Observation: Which leaf arrangement describes the leaves you are observing?

<u>Leaf Arrangements</u>

Alternate

Opposite

Whorled

Basal Rosette

Step Four - Leaves can be attached to the stem in four (4) ways: stalked, stalkless, clasping and perfoliate (wrapped around the stem).

Observation: How are the leaves attached to the stem of the plant you are observing?

Leaf Attachments

Stalked

Stalkless/Clasping

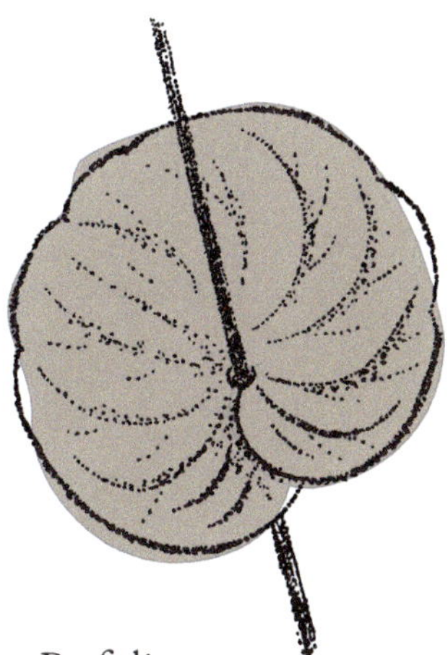

Perfoliate

Step Five - There are five (5) basic types of leaf edges: smooth, wavy, toothed, jagged, and lobed/deeply divided.

Observation: Which type of leaf edge does the plant you are observing have?

Leaf Edges

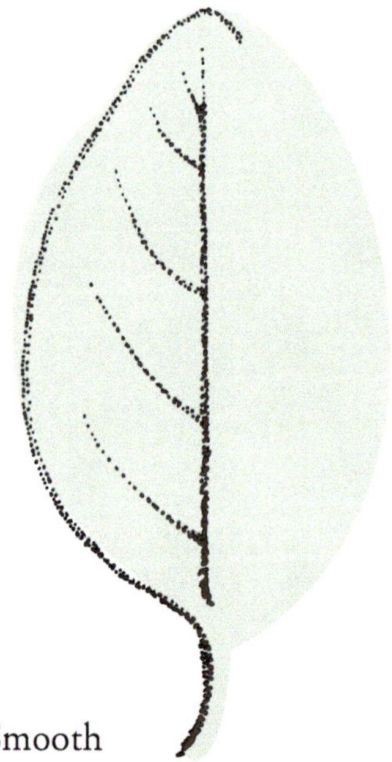

Smooth

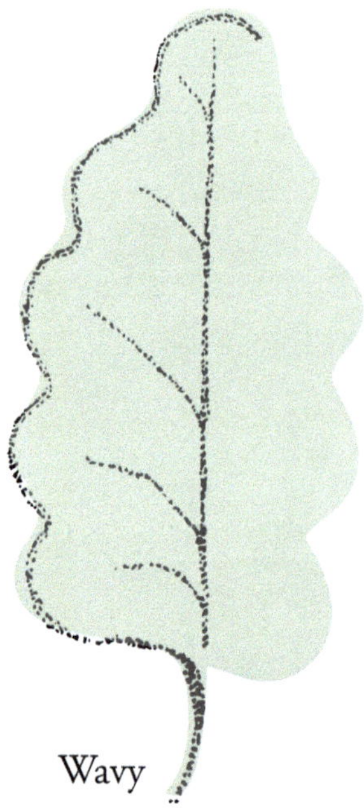

Wavy

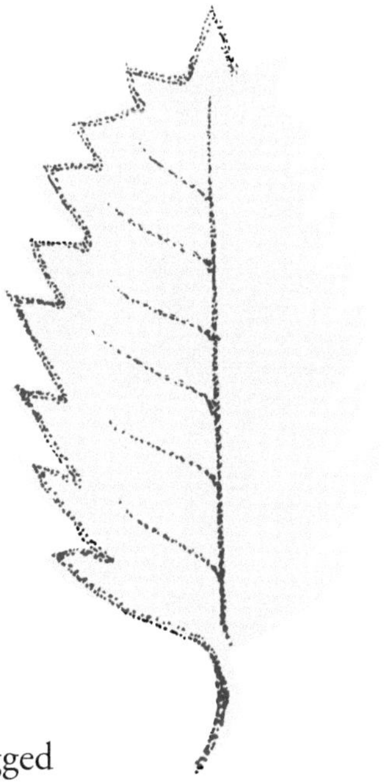

Jagged

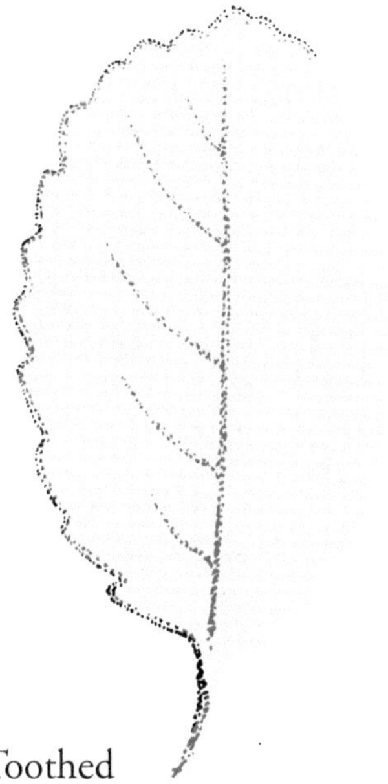

Toothed

Lobed/Deeply Divided

Step Six - Plant stems can be many shapes. The four (4) most common stem shapes are round, square, concave, or winged.

Observation: What is the shape of the plant stem you are observing?

Leaf & Plant Stem Parts

BRACT - a modified leaf located beneath the flower, not always present, sometimes mistaken for sepals

BUD - an undeveloped or unopened flower

LEAF AXIL - the joint where the leaf meets the stem

LEAF BLADE - the flat part of a leaf

LEAF EDGE - the outer margin of a leaf

LEAF STALK - the part of a leaf that attaches to the plant stem

PLANT STEM - the main part of a plant from which the flower stalk grows

STIPULE - a leaf–like structure at the base of a leaf stalk

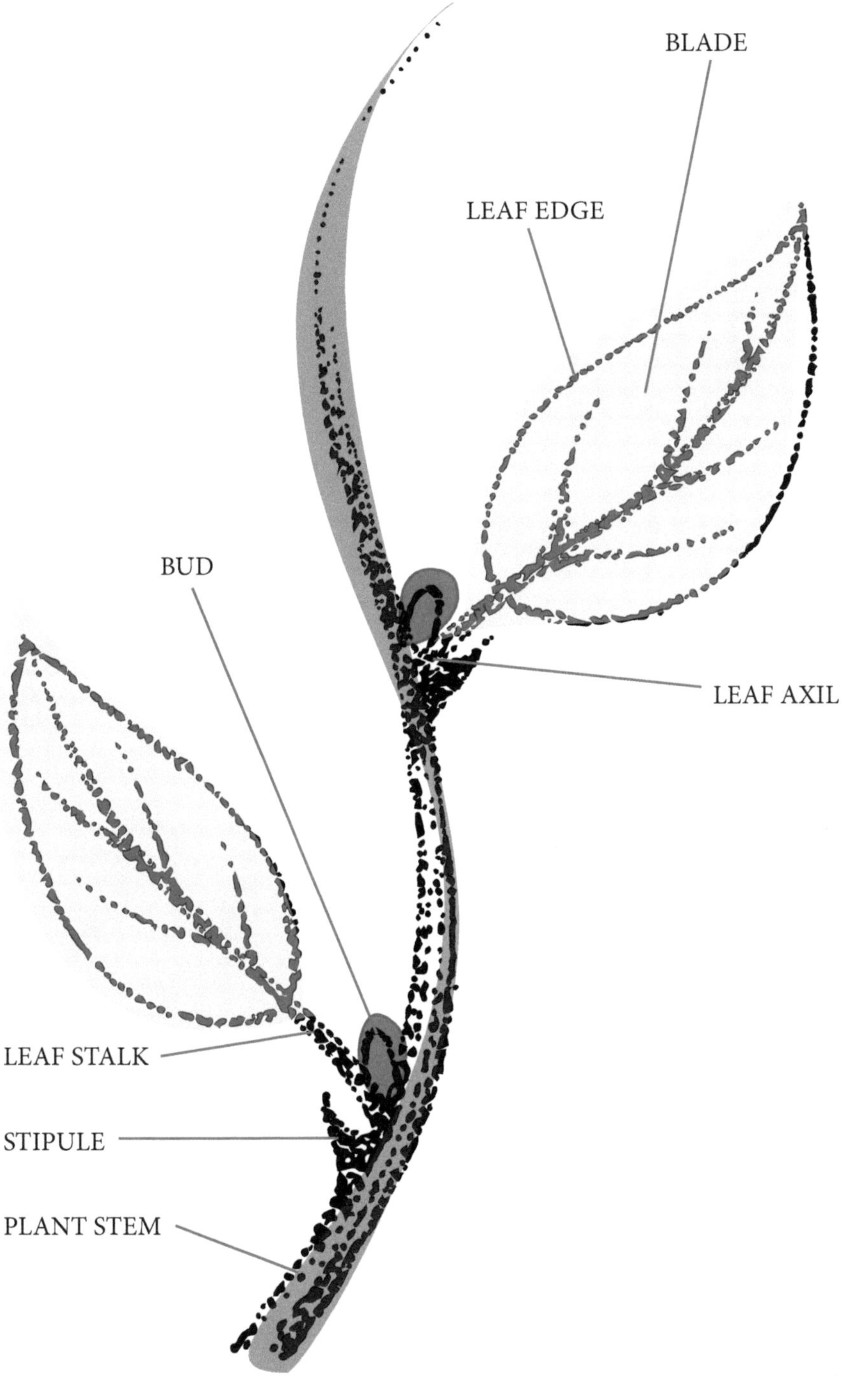
BLADE
LEAF EDGE
BUD
LEAF AXIL
LEAF STALK
STIPULE
PLANT STEM

CHAPTER THREE
ENJOYING WETLAND WALKS

WETLAND WALKS THROUGH THE SEASONS

Welcome to the wonderful world of wetlands! You'll love spending hours exploring these amazing places. Whether you visit in the spring, early summer or late summer, you'll find something new.

Wildflowers begin to bloom in the wetlands in early spring (April-May). Some don't start blooming until early summer (June- early July). Others continue blooming into late summer (Late July-September). By visiting the wetlands each season, you can learn to identify more than 100 common wetland flowers in the Great Lakes Watershed.

First, look through the plants listed for the season.

Spring	**(April – May)**
Early Summer	**(June – July)**
Late Summer	**(August – September)**

Second, keep in mind that plants may bloom for a shorter or longer time depending on the amount of rainfall and air temperature. So…if you don't find the plant you are observing in one season, try looking in the earlier or later season.

Enjoy walking through the wetlands in each season as you get to know the wonderful wetland plants of the Great Lakes Watershed.

SPRING WALK (April - May)

The leathery leaves of many early spring flowers push up their stems through the snow in early spring. If you hit the trails in April, you may see these earliest of wetland flowers: trailing arbutus and leatherleaf. Bright yellow marsh marigolds dot the edges of the ditches, creeks and swamps. They welcome the arrival of the early migrants such as the rose-breasted grosbeaks. Check for other spring wetland plants from April into the month of May.

White Flowers

Leatherleaf (shrub)
Pussy willow (shrub)

Pink Flowers

Bog rosemary (shrub)
Trailing arbutus (shrub)

Yellow Flowers

Marsh marigold

Green Flowers

Jack-in-the-pulpit

Leatherleaf

Family: Heath (Ericaceae)

Scientific Name: *Chamaedaphne calyculata* Height: 3' (0.9m)

FLOWER: Many, tiny (¼-½"), regular, stalked, five lobed, fused into urn shape white flowers dangle from small-leaved, terminal branches. Flowers hang from leaf axils like little lanterns.

LEAF: Simple, alternate, short-stalked, oval, (1-2") leaves; upper surface brown-green, leathery; lower, pale with small rusty dots; edges smooth or with small round teeth. Leaves smallest on flowering branches.

STEM: Older stems gray, shedding to show red inner bark.

NOTE: Evergreen shrub, 3' high. Same family as bog rosemary and wintergreen. Compare urn-shape flowers and evergreen characteristics.

Pussy willow Family: Willow (Salicaceae)

Scientific Name: *Salix discolor* Height: 60" (1.5m)

FLOWER: Catkins appear and mature before leaves in spring, stalkless, 1" long, sometimes with 2-3 small leaves at the base.

LEAF: Simple, oval and short-tapered to tip, dark green, smooth above, underside is red-hairy when young, becoming white-waxy, smooth and not wrinkled, margins entire or with few rounded teeth.

STEM: Yellow-brown to red-brown, dull, smooth with age or with patches of fine hairs.

NOTE: Produce soft, silver tufts in the spring that resemble tiny cat toes.

Bog Rosemary Family: Heath (Ericaceae)

Scientific Name: *Andromeda glaucophylla* Height: 4-30" (10-76cm)

FLOWER: Small clusters of 4-5 tiny, regular, 5-lobed, urn-shaped, white to pink flowers droop from ends of arching branches.

LEAF: Simple, alternate, tapered to base, bluish-green, narrow leaves carry sharp spine at tip. Surface leathery, with inward rolling edges; underside covered with woolly, white hairs.

STEM: Gray to black; twigs brown with linear hairs or smooth.

NOTE: Evergreen shrub. Compare underside of leaf to Labrador tea leaf, which shares evergreen, leathery, rolled-under leaf with woolly hairs. Traits of some heath plants.

Trailing Arbutus

Family: Heath (Ericaceae)

Scientific Name: *Epigaea repens*

Height: 1-6" (2.5-15cm)

FLOWER: Tiny (¼-½"), regular, five petals, white to pink, fragrant flowers arranged in compact clusters at ends of stems; flaring petals fuse into a short tube at their base

LEAF: Simple, alternate, stalked, oval, dull green leaves (1-3") with edges lined by stiff rust-brown hairs

STEM: Woody, creeping, covered with fine hairs

NOTE: Very early bloomer; leaves often hidden by leaf litter and sometimes, snow. "Epigaea" means tight to the earth - an earth hugger!

Marsh Marigold Family: Buttercup (Ranunculaceae)

Scientific Name: *Caltha palustris* Height: 4"-24" (10-60cm)

FLOWER: Regular, medium, not petals but 4-9 petal-like sepals which are green in a bud and become a cheery bright yellow flower. Flowers form at the end of stems and upper leaf axils to form a loose cluster.

LEAF: There are 2 types of simple leaves: numerous, long-stalked, large kidney or heart-shaped basal leaves and smaller, short-stalked or stalkless alternate leaves on flowering stems. Leaves are bright green with smooth or barely toothed edges.

STEM: Smooth, long, hollow; some upright, some relaxed, forming low rounded mounds.

NOTE: Species said to have more than 25 common names; one of them is "cowslip." The stem and leaves are edible when cooked, but poisonous when raw. Native Americans used the boiled, mashed roots for treating running sores.

Jack-in-the-pulpit

Family: Arum (Araceae)

Scientific Name: *Arisaema triphyllum* Height: 1-3' (0.3-0.9m)

FLOWER: Irregular, green/purple striped pulpit and canopy (spathe) covers and houses the club-like Jack (spadix). The reproductive all-male or all-female flowers grow at the base of the club, which in female flowers produce a cluster of red berries.

LEAF: Compound, divided into 3 stalkless leaflets, one or two compound leaves per plant, rising above the flower. Leaflets are dull green above and below with slightly wavy edges. If you look hard, you'll see a groove inside the edge that follows the whole margin, making a little border on each leaflet.

STEM: Stout, smooth, often mottled green/purple near base.

NOTE: Can be confused with trillium, which also has three-part leaves.

SPRING WALK CHECKLIST

- O Bog rosemary (shrub)
- O Jack-in-the-pulpit
- O Leatherleaf (shrub)
- O Marsh marigold
- O Pussy willow (shrub)
- O Trailing arbutus (shrub)

FIELD NOTES

FIELD NOTES

FIELD NOTES

EARLY SUMMER WALK (June - Early July)

You know that summer has finally come when the sun has warmed the surface of the water, there are no more frosts, and soil is ready for planting. Early summer flowers start to bloom in June, and once summer is in full swing, these wetland wildflowers can be seen through the month of July. Be sure to take your insect repellent, because bugs're hatching!

White Flowers

Bunchberry
Buttonbush (shrub)
Canada mayflower
Clintonia
Common wintergreen (shrub)
Creeping snowberry (shrub)
Goldthread
Labrador tea (shrub)
Purple meadow rue
Rough bedstraw
Starflower

Red Flowers

Northern pitcher plant

Pink Flowers

Calypso orchid
Dragon's mouth orchid
Fringed polygala/gaywings
Grass pink
Pink lady's slipper
Rose pogonia
Showy lady's slipper
Small cranberry (shrub)
Swamp milkweed
Swamp rose (shrub)
Twinflower (vine)

Purple Flowers

Marsh thistle/European swamp thistle (**Invasive**)
Purple avens
Purple loosestrife (**Invasive**)

Yellow Flowers

Golden ragwort
Large yellow lady's slipper
Small yellow lady's slipper
Northern swamp buttercup
Shrubby cinquefoil (shrub)
Smooth loosestrife
Tufted loosestrife
Yellow flag iris (**Invasive**)

Orange

Wood lily

Blue Flowers

Blue flag iris

Green Flowers

Nude miterwort/Bishop's cap

Bunchberry Family: Dogwood (Cornaceae)

Scientific Name: *Cornus alternifolia* Height: 3-8" (7.5-20cm)

FLOWER: This flower is deceptive. It looks like a flower with four petals, but actually the four white parts are bracts that probably attract pollinators. The bracts surround the tiny, creamy or yellow-green regular flowers. The flowers fruit into a bunch of red berries, hence the name Bunchberry.

LEAF: Whorl of 4-6 simple, stalkless, oval leaves tapered at both ends with deeply parallel veins and smooth edges.

STEM: Green, smooth, round, erect, woody.

NOTE: Grow in colonies. Flower structure very similar to its relative, the flowering dogwood tree. Although this flower looks like a composite flower, it lacks the involucre of bracts, a distinguishing hallmark of the composite family. Host plant for spring azure butterfly caterpillars.

Buttonbush Family: Madder (Rubiaceae)

Scientific Name: *Cephalanthus occidentalis* Height: 3-12' (0.9-3.5m)

FLOWER: Tiny, tubular, 4-lobed, creamy-white flowers clustered into bell-shaped, stalked heads, attached at stem ends and leaf axils. Flower styles with enlarged stigma protrude well beyond the flower edge to create a halo effect.

LEAF: Simple, opposite or whorls of 3, stalked, oval, edge smooth or wavy.

STEM: Woody, stems green-brown with lighter lenticels; older grey-brown

NOTE: Grows as a shrub along rivers and streams.

Canada Mayflower

Family: Lily (Liliaceae)

Scientific Name: *Maianthemum canadense* Height: 2-6" (5-15cm)

FLOWER: Looking close-up, this small white flower has a star-burst appearance; 2 petals and 2 petal-like sepals make it look 4-petaled; 4 pollen-laden stamen extend beyond petals creating the "burst." All flowers are stalked and clustered on a short spike and rise above the leaves.

LEAF: Simple, alternate, stalkless, oval with heart-shape base; edge smooth. 2-3 leaves on blooming plants; 1 leaf plants do not have flowers.

STEM: Zig-zag between leaves; green, round, smooth.

NOTE: Also called wild lily-of-the-valley; reddish berries eaten by grouse and small mammals in wet and dry habitats.

Clintonia

Family: Lily (Liliaceae)

Scientific Name: *Clintonia borealis* Height: 5–10" (12-25cm)

FLOWER: Regular, small, yellowish–green, drooping, bell–shaped flowers with 3 petal–like sepals and 3 petals that make it look like 6 petals. 2–8 flowers borne in loose cluster at top of leafless stalk

LEAF: Basal leaves only, 2–4 shiny dark green elliptical or oblong leaves, thick, edges finely hairy

STEM: Only stem is flower stalk, green, round, downy, leafless

NOTE: Flowers turn into fruit as pure blue, oval berries, held high on the stalk through summer; thus one of its names is Blue Bead Lily. Warning: Berries are poisonous!

Wintergreen Family: Heath (Ericaceae)

Scientific Name: *Gaultheria procumbens* Height: 2-6" (5-15cm)

FLOWER: Small, white to pink, regular, fine–toothed, fused, urn–shaped flower, each flower stalked and dangling from leaf axils; 1–5 flowers per plant

LEAF: Simple, alternate, oval to round, short stalked, edge with few low bristle-tipped teeth. New leaves light green or reddish and tender. Later leaves thick, shiny, hard, smooth and evergreen.

STEM: Round, green/brown, few hairs or bristles

NOTE: Delicious wintergreen aroma; used to flavor teas, candies and medicine

Creeping Snowberry Family: Heath (Ericaceae)

Scientific Name: *Gaultheria hispidula* Height: 1-2" (2.5-5cm)

FLOWER: A diminutive fairy garden size plant with tiny, regular, 4-lobed, bell-shape white flowers on curved stalks that grow from the leaf axils; flowers often hidden by leaves (inset).

LEAF: Simple, opposite, oval, very small; all leaves same size, arranged on both sides of stem. Leaves are short-stalked with edges rolled under, dark green on top, paler below with bristly hairs.

STEM: Delicate, hairy, intertwining stems creep over surface of old logs, creating a lovely embroidery.

NOTE: Evergreen, low-growing shrub. White berries slightly wintergreen flavored.

Goldthread

Family: Buttercup (Ranunculaceae)

Scientific Name: *Coptis groenlandica* Height: 3-6" (7.5-15cm)

FLOWER: Regular, 4-7 white, petal-like sepals encircle 15-25 small but noticeable white stamens, petals absent. In the center are 3-7 pale yellow-green stalked pistils with elongated styles that frequently hook downward. Each flower grows on a separate thin stalk rising from the ground and reaching above the leaves.

LEAF: Compound, 3 palmate leaflets which rise on their own separate basal stalk. They are bristle-tipped, and obscurely lobed with rounded teeth. Evergreen leaves are conspicuously shiny with rich green color.

STEM: All leaves and flowers have separate, thin, wiry stalks from the ground. There is no main plant stem.

NOTE: Roots are bright yellow color. Native Americans boiled them to make a dye or to treat sore gums and mouth sores.

Labrador Tea Family: Heath (Ericaceae)

Scientific Name: *Chamaedaphne calyculata* Height: 3' (0.9m)

FLOWER: Regular, 5-petaled, small white flowers with 5-7 stamens protruding beyond petals and borne in dense but delicate small terminal clusters.

LEAF: Simple, alternate, narrowly oval to oblong, round-tipped, evergreen. Leaves are thick and leathery with smooth upper surface, rolled edges. Underside covered with thick, cottony hairs that are white when new and turn rust color with age.

STEM: Gray or red brown, twigs covered with woolly, curly brown hairs.

NOTE: This small evergreen shrub seems dressed to withstand the harshest winters with its woolly, leathery coverings. Native Americans and colonists during the American Revolution used the leaves for tea.

Purple Meadow Rue Family: Buttercup (Ranunculaceae)

Scientific Name: *Thalictrum dasycarpum* Height: 4-8' (1-2.5m)

FLOWER: Regular, medium; male and female flowers grow on separate plants; both male and female plants have flowers on upper half of plant in large, branching clusters; flowers have no petals, but petal-like sepals early in blooming period. Male plants' flower clusters are comprised of dangling string-like filaments ending with yellow, pollen-laden anthers blowing in the wind. Female plants have more rigid flowers with clusters of pistils and ovary-style stigma pointing outward.

LEAF: Branched-compound; 20-30 leaflets per leaf found mostly on lower half of plant; stalked leaflets are rounded at base and usually 3-lobed at tip; leaves on upper flowering branches are stalkless. Top of leaves are blue-green, paler underneath, not waxy.

STEM: Tall, robust, most often purple.

NOTE: Native Americans used the plant to reduce fevers and improve blood circulation.

Rough Bedstraw Family: Bedstraw (Rubiaceae)

Scientific Name: *Galium asperellum* Height: 8-36" (20-91cm)

FLOWER: Regular, tiny, unimpressive, 4-petaled flowers in spreading few-flowered clusters at stem ends and upper leaf axils.

LEAF: Simple, in whorls of 6 stalkless leaves (sometimes 5); narrowly oval ending in sharp tip; edges and mid-vein lined with rough hairs on underside. Leaves are small (½"- ¾").

STEM: Square shape with downward pointing, clinging hairs on stem edges. New stems branch from leaf axils at 45 degree angle. Stems are weak, spreading, reclining, gaining support from surrounding plants.

NOTE: Stems grab like Velcro on contact.

Starflower Family: Primrose (Myrsinaceae)

Scientific Name: *Trientalis borealis* Height: 4-8" (10-20cm)

FLOWER: Regular, small white star-shaped flower with 5-9 (usually 7), pointed, slightly overlapping petals; very delicate; 1-2 flowers on separate, wiry stalks that rise an inch or so above leaves. Flower has stamens, same number as petals, with white filaments carrying the anthers; yellow-green pistil with elongated style in center of flower.

LEAF: Simple. Lance-shaped, smooth, group of 5-7 stalkless leaves attached to central stem in a whorl; leaves 2-4", not all the same size gives plant an uneven look; few scale-like leaves below whorl.

STEM: Green, smooth, unbranching, bare except for scale-like leaves supporting whorl of leaves and flowers.

NOTE: Once you see this flower, you'll always know it. Even without its star-like flower, the leaves suggest a star-shape to give you a clue.

Northern Pitcher Plant Family: Pitcher Plant (Sarraceniaceae)

Scientific Name: *Sarracina purpurea* Height: 8-24" (20-61cm)

FLOWER: Large (2–3") regular, wine–red, globular bell–shaped on separate nodding leafless stalk; five large petal–like sepals surround inner circle of five petals, all curving inward closing the ovary and stamens. Both stalk and flower are sturdy, persisting through summer.

LEAF: Simple, basal, clumped leaves grow upright from base of plant; hollow winged tube with flared lip that holds rain water; smooth on the outside; inside of lip has downward pointing hairs; leaf mainly green, often with dark red veining. Although the leaf is "simple," it has a very complex structure (inset).

STEM: No plant stem; just individual flower and leaf stems.

NOTE: Carnivorous– Insects are attracted by reddish color and scent of the plant. They are trapped by the downward pointing hairs that prevent their escape from the tube–like leaves. Then enzymes digest the insect, which becomes the source of nitrogen and other nutrients for the plant.

Calypso Orchid

Family: Orchid (Orchidaceae)

Scientific Name: *Calypso bulbosa var. americana*

Height: 2-9" (5-23cm)

Status: Threatened in Michigan & Wisconsin

FLOWER: Small, single flower at end of stem, 3 pink erect sepals, 2 lateral petals, white column, scoop-shaped less than 1" with dark maroon marks, white lip, spoon-shaped with 2 horns below and yellow beard above

LEAF: Single, basal leaf 1-2" long (2-6cm), appears in late summer and withers by spring

STEM: Purple

NOTE: Needs pollination from 10 species of bumble bees; lives no more than 5 years, may be dormant for several years after blooming, intolerant of soil temperatures above 59 degrees F (15 C).

Dragon's Mouth Orchid Family: Orchid (Orchidaceae)

Scientific Name: *Arethusa bulbosa* Height: 3-16" (8-41cm)

Status: rare, protected, special concern in Wisconsin

FLOWER: Large, rose purple-brilliant pink single flowers, 2 pink sepals stand erect like horns of a dragon, 2 fused petals above pale pink column, wavy & downward arching lip has darker magenta, white and yellow marks and fleshy "hairs"

LEAF: Single, grass-like leaf 5-10" long (12-25 cm), appears after bloom fades

STEM: Green

NOTE: Susceptible to frost, so may not produce seeds every year. Reproduces asexually by spreading slowly from tubers underground.

Fringed Polygala/Gaywings

Family: Milkwort (Polygalaceae)

Scientific Name: *Polygala paucifolia* Height: 3-6" (7.5-15cm)

FLOWER: Irregular, small, pink-purple flower that looks like an airplane; two flaring lateral sepals create the wings while the 3 petals fuse into a small tube, the lower tipped with bushy fringe, resembling the body and propeller. Truly a joy to come upon.

LEAF: Evergreen, smooth, alternate, low growing; stalked leaves cluster near top of stem; flowers grow from leaf axils

STEM: Green, low, smooth, spreading close to the ground

NOTE: At first glance Gaywings may look like small orchids, but there is no relationship.

Grass Pink

Family: Orchid (Orchidaceae)

Scientific Name: *Calopogon tuberosus var. tuberosus*

Height: 4-22" (10-56cm)

Status: protected

FLOWER: Small, sweet-smelling, 2-15, bright pink to pale purple, 1-2" (2.5-5 cm) in loose showy cluster, sepals and petals bright pink and spreading, small column, anvil-shaped, upright with yellow-tipped bristles, small lip, narrow, curves downward

LEAF: Erect, grass-like leaf, strongly ribbed with parallel veins, 2-20" long

STEM: Strong, tall single green stem

NOTE: Pollinated by small bees.

Pink Lady's Slipper Family: Orchid (Orchidaceae)

Scientific Name: *Cypridium acaule* Height: 5-22" (13-56 cm)

Status: Protected

FLOWER: Small, single, pink, nodding, greenish-yellow to reddish-brown sepals and petals, sepals arch over pouch opening, petals twisting outward, small, pink column, large inflated lip, deep pink to magenta with darker, reddish veins, up to 2.5" (6 cm) long; top is split with edges rolled inward

LEAF: 2 basal leaves clasping the stem, up to 11"

STEM: Strong, green, clumps of 10 or more

NOTE: Pollinated by bumble bees. Also reproduces by spreading rhizomes. Slow growing. Seed to first bloom takes 8-10 years.

Rose Pogonia Family: Orchid (Orchidaceae)

Scientific Name: *Pogonia ophioglossoides* Height: 6-20" (15-50 cm)

Status: Protected

FLOWER: One pink to purple flower at top of stem, pink sepals, spreading to form hood over column, pink petals, smaller than sepals, small column, inconspicuous, small lip, starts narrow with dark, magenta marks and broadens toward the fringed tip, center has yellow, white or purple fleshy hair-like bristles, curves downward

LEAF: 1 erect, green, pointed oval leaf sheaths midway up the stem, 4" (10 cm) long, deep veins, fleshy

STEM: Strong, tall single green stem

NOTE: Sends out shoots from underground roots to form new plants.

Showy Lady's Slipper

Family: Orchid (Orchidaceae)

Scientific Name: *Cypridium reginae*

Height: 8-36" (20-91cm)

Status: Protected

FLOWER: Large, nodding, 1-2, pink and white, white sepals, arched over column, white petals, small white column, large inflated lip, white with pink to deep fuchsia, up to 2.2" long; top is circular with edges rolled inward; inside has purple spots & lines

LEAF: 3-12 pleated, pointed oval leaves with fine hairs alternate up stem, deeply veined, up to 10" (26 cm) long

STEM: Strong, green, clumps of 25 or more

NOTE: Pollinated by bees. Also reproduces by spreading rhizomes.

Small Cranberry

Family: Heath (Ericaceae)

Scientific Name: *Vaccinium oxycoccus*

Height: 4-8" (10-20cm)

FLOWER: Regular, pale–pink, 4 lobed, lobes curved back forming a Turk's cap, stamens form a beak; flower stalk grows from the tip of branch, ½–1 ½" tall with 2 red bracts borne at or below the middle of the stalk. Each stalk has 1-4 flowers.

LEAF: Simple, alternate, tiny (1⁄16–¼" long), narrowly pointed and ovate; leathery; top green and white below, edges rolled under

STEM: 1 ½" stems horizontally trail along ground, from them grow upright branches, the tips of which bear flower stalks 4–6" high

NOTE: The distinguishing characteristics that separate this from large cranberry are size, flower stalks growing from tips of branches, and pointed leaves with white underneath.

Swamp Milkweed Family: Milkweed (Asclepiadaceae)

Scientific Name: *Asclepias incarnata* Height: 3-4' (0.9-1.2m)

FLOWER: Regular with petal-like hoods, each with an awl-shaped "horn" projecting from the opening, pink to purple-red, numerous umbels at ends of stems and from upper leaf axils.

LEAF: Simple, opposite, mostly lance-shaped, tapered to a sharp tip, margins entire.

STEM: Stout, branched above, smooth except for short hairs on upper stem.

NOTE: Native Americans used the plant as a diuretic and to treat lung problems.

Swamp Rose

Family: Rose (Rosaceae)

Scientific Name: *Rosa palustris*

Height: 3–6' (0.9-1.8m)

FLOWER: Regular, pink, medium size, five–petaled; flowers borne singly or in small clusters on leafy branches; flower stalks and sepals have stalked glands.

LEAF: Alternate, pinnately compound, usually with 7 leaflets; leaflets oval, longer than wide, edges finely toothed, underside mid–rib often softly hairy; leaf stems have narrow–winged stipules at base

STEM: Red–brown, bushy with paired downward curved thorns at each leaf node; smooth between

NOTE: Old fashioned–style rose; fruit is red–orange round "hip"

Twinflower

Family: Honeysuckle (Caprifoliaceae)

Scientific Name: *Linnaea borealis*

Height: 3–6" (7.5-15cm)

FLOWER: Regular, small, pink, 5–lobed bell–shaped fragrant flowers, held in nodding pairs by a Y–shaped stalk, looking like a tiny, lamppost with 2 lamps. The 3–6" stalk arises from a creeping stem.

LEAF: Simple, opposite, evergreen, short–stalked, round or oval, somewhat 3–lobed at tip or smooth, hairy, edges rolled under. Leaves paired low along stem and at base of flower stalks.

STEM: Slightly woody trailing vine with numerous erect leafy branches rising at intervals which bear the flower stalks

NOTE: Twinflower's genus is named after Carl Linnaeus, a Swedish botanist (1707–1778). He laid the foundation for the scientific system for naming plants and animals with two words, genus and species.

European Marsh Thistle Family: Aster (Asteraceae)

Scientific Name: *Cirsium palustre* Height: 1.5–5' (0.5-1.5m)

FLOWER: Composite flower heads, purple disk flowers only; short–stalked, involucre bracts not spiny; flowers in groups of 3 at end of stem

LEAF: Simple, deeply lobed, edges spine–toothed; leaf tapered at base then joins stem as spiny wing. Lower leaves have long hairs and, to some degree, have loosely matted hairs

STEM: Tall, spiny green

NOTE: INVASIVE- May grow 7 feet (2 meters) high. Biennial. Basal florets year one. Flowers year two. Spreads aggressively.

Purple Avens Family: Rose (Rosaceae)

Scientific Name: *Geum rivale* Height: 1–2' (30-60cm)

FLOWER: Regular, small, nodding globular flowers with 5 inconspicuous purple infused yellow petals; flowers in groups of 3 on upper stems. As flower matures, petals disappear and styles attached to developing seeds grow plume–like and noticeable.

LEAF: Low, basal leaves pinnately compound, stalked, coarsely toothed, upper 3 leaflets much larger than the rest; stem leaves are 3–lobed, stalked or stalkless, coarsely toothed with green or purplish stipules

STEM: Hairy, largely unbranched, erect

NOTE: Pollinated by bees, less often by flies and beetles. Seeds are burr-like and get stuck in coats of small mammals. Called Indian Chocolate. Used to treat digestive disorders.

Purpose Loosestrife Family: Loosestrife (Lythraceae)

Scientific Name: *Lythrum salicaria* Height: 2–5' (0.6-1.5m

FLOWER: Regular, small, purple–magenta mainly 6–petaled, flowers clustered, on long, showy leafy spikes that rise from leaf axils of upper plant stems, often 2 or more spikes per stem. Sepals of flowers join to make elongated, hairy tube from which 12 joined stamens grow to varying lengths.

LEAF: Simple, opposite or whorls of 3, narrow, lance–shaped, clasping, heart–shaped at base, size reduced within spike and may become alternate. Leaf edge smooth, surface downy.

STEM: Erect, tangled, many per plant

NOTE: INVASIVE. Spreads and forms dense colonies. Reproduces in three ways: roots, seeds, or pieces of the plant that break off and establish a new plant.

Golden Ragwort Family: Aster (Asteraceae)

Scientific Name: *Packera aurea* Height: 6-30" (15-76cm)

FLOWER: Aster-like, composite, forms heads of 8-15 small, yellow-rayed flowers, involucre has linear, pointed, green bracts. Flowers grow in a branched terminal cluster.

LEAF: Basal leaves that grow at base of plant, not from stem, are large, roundish, blunt-toothed with heart-shaped base, long stalks, purplish below. Stem leaves are stalked, long oval with rounded, toothed tip and narrow deep lobes at base; upper leaves are narrow, pinnately toothed, unstalked.

STEM: Green, round, slightly hairy.

NOTE: The Cinnabar moth feeds on golden ragwort as its host plant. WARNING - May cause liver damage.

Large Yellow Lady's Slipper

Family: Orchid (Orchidaceae)

Scientific Name: *Cypridium parviflorum var. pubescens*

Height: 7-30" (18-76cm)

FLOWER: Large, single, yellow, nodding. Sepals and petals greenish-yellow marked with purplish-brown spots or stripes. Sepals arch over pouch opening. Lip is an inflated, deep yellow pouch with reddish marks inside.

LEAF: 3-6 oval-shaped, clasp stem, bright green, ribbed, 8" (20 cm).

STEM: Strong, green

NOTE: Pollinated by small bees; spreads by rhizomes underground.

Small Yellow Lady's Slipper

Family: Orchid (Orchidaceae)

Scientific Name: *Cypridium parviflorum var. makasin*

Height: 6-17” (15-43cm)

Status: protected

FLOWER: Very small, single, yellow, nodding; sepals arch over pouch opening; petals dark reddish-brown to purple, curling; column small, yellow, lip an inflated, deep yellow pouch with reddish-purple marks inside

LEAF: 3-6 oval-shaped, alternately clasping stem, bright green and ribbed, 7” (18 cm)

STEM: Strong, green

NOTE: Pollinated by small bees; spreads by rhizomes underground.

Northern Swamp Buttercup

Family: Buttercup (Ranunculaceae)

Scientific Name: *Ranunculus hispidis* Height: 1-3' (30-91cm)

FLOWER: Regular, small, usually 5–petaled with reflexed green sepals, bright, shiny open–face flower which holds a ring of stamens surrounding numerous green pistils. Several flowers are held above leaves on sparsely branched stems.

LEAF: Compound, alternate, varied. Some leaf stems grow from the ground and are called basal leaves. They are larger; longer stalked; divided into 3–stalked leaflets, the center one longer stalked; each leaflet is also 3–lobed and deeply cut. The leaves on flower stems are smaller, alternate and 3–parted with jagged edges. Leaves may have hairs on veins.

STEM: Hollow, coarse, may be hairy; ascending or trailing along the ground.

NOTE: Looking into the flower at the base of the petals, notice a change in color to translucent yellow–green with fine dark green lines that are nectar guides for insects.

Shrubby Cinquefoil

Family: Rose (Rosaceae)

Scientific Name: *Dasiphora fruticosa* Height: 2-3' (61-91cm)

FLOWER: Regular, small, usually 5–petaled, bright yellow, rose-like; one or fewer flowers at the ends of branches; 5 sepals are joined at the base, deeply lobed and alternate with 5 lance–shaped bracts

LEAF: Pinnately compound, alternate, 3–7 (usually 5) leaflets, small oval to oblong; tapered at both ends; edges smooth to rolled under, short–stalked, upper–side deep green, underside paler with silky hairs.

STEM: Many branches, bushy shrub

NOTE: Native Americans used the leaves as an astringent and the roots to treat indigestion.

Smooth Loosestrife Family: Primrose (Primulaceae)

Scientific Name: *Lysimachia quadriflora* Height: 1–3' (30-91cm)

FLOWER: Regular, 1", 5–petaled yellow flowers in upper axils on slender stalks, single or in whorls of 4, slightly toothed petals, slightly nodding.

LEAF: Simple, opposite, stalkless, narrowly lance–shape; edges smooth, stiff.

STEM: Single stem is unbranched or slightly branched below, erect, smooth, round.

NOTE: Sometimes confused with other loosestrife species, smooth loosestrife is distingued by its narrow linear leaves.

Tufted Loosestrife   Family: Primrose (Primulaceae)

Scientific Name: *Lysimachia thyrsiflora*   Height: 1–3' (30-91cm)

FLOWER: Regular, tiny, mostly 6–lobed, densely clustered in stalked oval heads growing from leaf axils mid–way up the stem, petals fused, showy stamen – twice as long as petals

LEAF: Simple, opposite, narrow, lance–shaped, stalkless, edges toothless, wavy; smooth upper surface, some hair along mid–rib beneath

STEM: Erect, usually unbranched, smooth or with patchy brown hairs, reddish lower stem, stems clustered

NOTE: Flowers in round clusters on hairy stems that arise from the leaf axilsabout midway up the plant. No flowers at the top of the plant.

Yellow Flag Iris

Family: Iris (Iridaceae)

Scientific Name: *Iris pseudocorus*

Height: 2-3' (61-91cm)

FLOWER: Irregular, large, yellow, complicated; several growing on each stiff stem. 3, narrow, erect yellow petals in center of flowers with 3 larger, spreading, round-tipped, curving sepals around the outside. The female part of the flower arches like a tongue part–way over the sepal, where bees enter.

LEAF: Long, narrow, smooth, sword–like, stiff–pointed blade. Leaves grow from underground rhizomes in flat clusters that create flat sprays of greenery.

STEM: Flower stems also grow from same rhizomes. They are round, smooth, green and about height of leaves with one or two branches at top.

NOTE: INVASIVE; toxic roots.

Wood Lily

Family: Lily (Liliaceae)

Scientific Name: *Lilium philadelphicum* Height: 2–3' (61-91cm)

FLOWER: Regular, large, erect, showy, orange, 6–parted, stalked flower with 3 large, petal–like sepals and 3 large petals, noticeably spotted inside with red or purple oblong dots near yellowed base. Flower open, upward pointing; 6 stamens with deep purple anthers surround blackish 3–lobed stigma. 1–5 flowers per plant.

LEAF: Simple, narrow lance–shaped lower leaves alternate; upper whorled with 5–7 stalkless leaves, edges smooth, parallel–veined

STEM: Stiff, erect, branched only at top; green, round.

NOTE: CAUTION - Very toxic to cats.

Blue Flag Iris Family: Iris (Iridaceae)

Scientific Name: *Iris versicolor* Height: 2–3' (61-91cm)

FLOWER: Irregular, large, showy blue–violet flowers that consist of 3 each of three floral parts. There are 3 purple, petal–like sepals that arch and spread with solid purple, rounded tips. The rest is white streaked with purple veins and a patch of yellow in the throat. There are 3 upright, erect, purple petals half as long as sepals. There are 3 purple strap–like branches of the style that over–arch the sepals. Several flowers are found at branched tip of stem.

LEAF: Simple, narrow, sword–like, firm, mostly from lower part of stem, clasping at base and shorter than stem.

STEM: Round, stiff, rising from thick, spreading rhizomes creating colonies. Stem branched at top of stem where flowers grow.

NOTE: WARNING - Rhizomes may be poisonous. Can cause dermatitis.

Nude Mitrewort

Family: Saxifrage (Saxifragaceae)

Scientific Name: *Mitella nuda*

Height: 4–10" (10-25cm)

FLOWER: Regular, tiny, yellowish–green, 5–petaled; 5–lobed, rounded or pointed joined sepals look flower–shaped; true petals are between sepal lobes, feathery or fishbone–like with 9–11 spines. 10 yellow-tipped stamens surround ovary at center. 3–12 stalked flowers are borne on a glandular hairy stem.

LEAF: Simple, mostly basal, roundly heart–shaped on long slender stalk; edges roundly toothed, leaf surface hairy.

STEM: Slender, stiff, usually leafless, but can have single leaf mid–stem; hairy, upper part glandular.

NOTE: Flowers fragile, present in early bloom but may not last long. Also known as Bishop's cap.

FIELD NOTES

EARLY SUMMER WALK CHECKLIST

- O Blue flag iris
- O Bunchberry
- O Buttonbush (shrub)
- O Calypso orchid
- O Canada mayflower
- O Clintonia
- O Common wintergreen (shrub)
- O Creeping snowberry (shrub)
- O Dragon's mouth orchid
- O Fringed polygala/gaywings
- O Golden ragwort
- O Goldthread
- O Grass pink
- O Labrador tea (shrub)
- O Large yellow lady's slipper
- O Marsh thistle/European swamp thistle (**Invasive**)
- O Northern pitcher plant
- O Northern swamp buttercup
- O Nude mitrewort/Bishop's cap
- O Pink lady's slipper
- O Purple avens
- O Purple loosestrife (**Invasive**)
- O Purple meadow-rue
- O Rose pogonia
- O Rough bedstraw

- O Showy lady's slipper
- O Shrubby cinquefoil (shrub)
- O Small cranberry (shrub)
- O Small yellow lady's slipper
- O Smooth loosestrife
- O Starflower
- O Swamp milkweed
- O Swamp rose (shrub)
- O Tufted loosestrife
- O Twinflower
- O Wood lily
- O Yellow flag iris (**Invasive**)

FIELD NOTES

FIELD NOTES

LATE SUMMER WALK (Late July-September)

Summer days are the warmest of the year, so more wetland wildflowers put on quite a show. Even as days start to get shorter and nights become cooler, you'll see many of these beauties into September.

White Flowers

- Boneset
- Grass of Parnassus
- Marsh bellflower
- Nodding ladies' tresses
- Tall white bog orchid
- Turtlehead

Red Flowers

- Cardinal flower
- Round-leaved sundew

Pink Flowers

- Spotted Joe-pye weed

Purple Flowers

- Swamp thistle
- Purple-fringed orchid
- Purple-stem aster

Yellow Flowers

- Bog goldenrod
- Rough-leaved goldenrod
- Smooth goldenrod

Orange Flowers

Jewelweed

Blue Flowers

Blue vervain

Brook lobelia

Great blue lobelia

Green

Northern green bog orchid

Boneset Family: Aster (Asteraceae)

Scientific Name: *Eupatorium perfoliatum* Height: 2–4' (0.6-1.2m)

FLOWER: Composite, white, disk flowers only, arranged in stalked, fuzzy–looking flat–top clusters at end of stems.

LEAF: Simple, opposite, lance–shaped, broad bases of leaves joined completely and pierced by stem; upper leaves sometimes separate, stalks absent. Leaf margins with rounded teeth and rough; upper margins with rounded teeth and rough; upper surface wrinkly, sparsely hairy; underside hairy.

STEM: Round, green, coarse, long–haired, erect.

NOTE: Native Americans used boneset to mend broken bones, reduce fever and treat constipation. May cause vomiting.

Grass of Parnassus

Family: Saxifrage (Saxifragaceae)

Scientific Name: *Parnassia glauca*

Height: 8–24" (20-61cm)

FLOWER: Regular, 5 white petals with striking translucent green veins; solitary flowers at top of tall, erect stalks, some leafless, some with very small, clasping leaf.

LEAF: Simple, growing from base of plant, parallel–veined, broadly oval with heart–shaped base. Edges and surfaces are smooth (inset).

STEM: Round, upright; leaf stems shorter and basal flower stems rise well above leaves.

NOTE: Used in herbal medicine to treat liverd disorders, indigestion and kidney stones. A very classy looking plant.

Marsh Bellflower Family: Bellflower (Campanulaceae)

Scientific Name: *Campanula aparinoides* Height: 8–24" (20-61cm)

FLOWER: Regular, tiny, 5–lobed, bell–shaped, color varying from light blue to lavender to white. Flowers are stalked and grow from upper leaf axils of stem. They are sparse and can get lost among leaves and reclining stem.

LEAF: Simple, alternate, stalkless, small, narrow, pointed tip; leaves smallest on upper stem; edges are smooth or fine–toothed, hairy along edge and midrib.

STEM: Slender, weak, reclining, 3–angled, slightly zig–zag with short, stiff hairs that grab like Velcro.

NOTE: Uses subterraneous runners to propogate, which concentrates plant in a small area. May not produce seeds. Sticks to other plants.

Nodding Ladies' Tresses Family: Orchid (Orchidaceae)

Scientific Name: *Spiranthes cernua* Height: 4-20" (10-50cm)

FLOWER: Small, 8-50, white, set in 2-4 tight, spiraling rows, tightly packed along top of stem; sepals white, forming an up-curved hood over the lid; petals white, curl up around small white column; lip small, white, arches downward, edges translucent & fringed edge looks like filigree.

LEAF: 2-6 narrow, blade-like leaves 3-11" (8-28 cm) long, longer leaves are basal, upper leaves reduced to pointed, clasping bracts

STEM: Strong, tall single green stem, covered with fine hairs

NOTE: Pollinated by bumble bees. Reproduces asexually in two ways: small plantlets and young plants from tips of rhizomes

Tall White Bog Orchid

Family: Orchid (Orchidaceae)

Scientific Name: *Platanthera dilatata var. dilatata*

Height: 10-40+" (25-102cm)

FLOWER: Small, 10-75, white flowers, tightly packed along stem; sepals white, arched over small greenish column with backwards curving green spur; petals cupped over column; lip small, dilated at base and tapered to blunt tip.

LEAF: 3-8 tapered, dark green blades, up to 10" (25 cm) long, sheath the stem. Spreading leaves become smaller as they climb the stem. Leaves form bracts at top of stem.

STEM: Tall, smooth, single green stem

NOTE: Pollinated by bees, butterflies and moths.

Turtlehead Family: Figwort (Scrophulariaceae)

Scientific Name: *Chelone glabra* Height: 1–3' (30-91cm)

FLOWER: Irregular, medium size, white or pink, 2–lipped cousin of the snapdragon, borne in a compact terminal spike–type cluster. The two lips are almost closed; the upper lip is arched over the bearded, lower lip.

LEAF: Simple, opposite, narrow, lance–shaped with pointed tip, stalkless or very short stalk; edges sharply toothed, leaf surface smooth, deeply veined and shiny.

STEM: Round, stout, upright, light green.

NOTE: Host plant for rare Baltimore butterfly.

Cardinal Flower Family: Bellflower (Campanulaceae)

Scientific Name: *Lobelia cardinalis* Height: 2–4' (0.6-1.2m)

FLOWER: Irregular, strikingly red, 2–lipped flowers; the upper lip with 2 erect petals, the lower lip with 3 spreading petals, all petals fused into tubed base. Stalked flowers are borne on a spike–like cluster, each sub–tended by a bract. Flowers bloom from bottom up.

LEAF: Simple, alternate, lance–shaped, upper leaves nearly clasping, lower leaves with short stalks.

STEM: Stiff, erect, usually unbranched, round, green.

NOTE: WARNING - Potentially toxic.

Round–leaved Sundew Family: Sundew (Droseraceae)

Scientific Name: *Drosera rotundifolia* Height: 2–4" (5-10cm)

FLOWER: Regular, tiny, 5 petaled, 2–20 white to pink flowers lined up in a raceme–like cluster at end of a nodding, leafless stalk. Buds open from bottom up, one at a time, only in full sun. The current bloom is located at the highest point above the point where the stalk turns downward (inset).

LEAF: Simple, basal, small, round, on stalk longer than leaf, which coils from apex in leaf bud stage. Upper leaf surface covered with long, glandular-tipped, reddish hairs that produce a sticky, glue–like substance.

STEM: Flowering stem leafless, leaf stems hairy.

NOTE: Native Americans used the plant as an anti-inflammatory and anti-spasmodic medicine. Insectivorous. Go to page 184 to read about insectivorous plants.

Spotted Joe–Pye Weed

Family: Aster (Asteraceae)

Scientific Name: *Eupatorium maculatum* Height: 2–10' (60–300 cm)

FLOWER: Composite, small pinkish–purple disk flowers only, 9–27 flowers clustered in small heads above a cylindrical involucre of overlapping bracts; small heads are grouped into large branched flat top terminal clusters, 5–10" wide

LEAF: Simple, whorls of 4–5 lance–shaped, stalked, coarsely forward–pointed toothed leaves; upper surface sparsely haired, more densely short–haired below

STEM: Round, tall, purple or purple–spotted; short–haired on upper branched stems of flower clusters

NOTE: Native Americans used roots for a diuretic and the plant to treat fever and inflammation.

Swamp Thistle

Family: Aster (Asteraceae)

Scientific Name: *Circium muticum*

Height: 2–6' (60–182 cm)

FLOWER: Composite, rose–purple disk flowers only; medium–size heads on end of branched upper stems, involucre sticky, cobwebby, not spiny

LEAF: Simple, alternate, deeply lobed with spiny tips along the edge, basal leaves stalkless; underside of leaves often with matted hairs

STEM: Stout, erect branched at top, long soft hairs when young, smoother with age

NOTE: Biennial plant that completes its life cycle on two growing seasons; typically, first year produces a basal rosette; second year produces flower bud, fruit and then dies

Small Purple Fringed Orchid Family: Orchid (Orchidaceae)

Scientific Name: *Platanthera psycodes* Height: up to 40" (100 cm)

FLOWER: Small, lilac to purple flowers, tightly packed at top of stem, blossoms open from top to bottom; purple sepals, arched over column; small, purple petals, curved upward; small, white column; small, purple lip, divided into 3 fan-shaped lobes, fringed at edge, tubular spur curves downward 1" (23 mm) long

LEAF: 4-12 pointed, ovate blades, up to 9" (23 cm) long, alternate low along the stem. Leaves sheath the stem and form bracts at top of stem.

STEM: Tall, smooth, single green stem

NOTE: Pollinated by butterflies by day and moths by night.

Purple–stem Aster

Family: Aster (Asteraceae)

Scientific Name: *Symphyotrichum puniceum* Height: 2-7' (0.6-2.1m)

FLOWER: Composite, large, ray flowers, blue or blue–violet (20–50), bracts are long, sometimes hairy, spreading to form an involucre

LEAF: Simple, alternate, lance–shape, stalkless, clasping, sharp–toothed, rough on top, long hairs below

STEM: Round, reddish–purple, bristly, unbranched below, some branching at top

NOTE: Variable (may have smooth toothless leaves and smooth green stems)

Bog Goldenrod

Family: Aster(Asteraceae)

Scientific Name: *Solidago uliginosa* Height: 2’-5’ (60-152 cm)

FLOWER: composite, yellow, clusters dense and narrow, form an erect narrow plume or wand

LEAF: simple, alternate, smooth and toothless or small teeth, lower and basal leaves are long (9”), tapering to a long stalk that partially sheathes the stem

STEM: round, smooth, hairless, erect, red to purple in color

NOTE: Native Americans used this goldenrod to wash wounds, cuts and burns. Can be used as a poultice on an open wound.

Rough–leaved Goldenrod

Family: Aster (Asteraceae)

Scientific Name: *Solidago patula*

Height: 2–7' (60–210 cm)

FLOWER: Composite, yellow small heads made of ray (6–12) flowers and disk flowers arranged on a series of spreading, arched, fountain–like branches at top of stem; heads are mostly on one side of branch

LEAF: Simple, alternate, basal leaves very large, pointed oval, base tapering to a long winged stalk (inset); pointed large teeth, edges forward, upper leaves smaller, stalkless, finely toothed or smooth; all leaves harshly rough above, smooth below

STEM: 4–sided, flower part smooth, upper short hairy

NOTE: Native Americans used the plant to treat nausea, dizziness, sunstroke and liver ailments.

Trail near Cedar Lake

Smooth Goldenrod Family: Aster (Asteraceae)

Scientific Name: *Solidago gigantea* Height: 3–7' (91–210 cm)

FLOWER: Composite, yellow heads made of ray (9–15) flowers and disk flowers; many heads cover one side of branches that form a plume–like cluster at top of stem

LEAF: Simple, alternate, lance–shaped, parallel–veined, tapered to a stalkless base, forward pointing teeth, top smooth, underside smooth or with short hairs on veins

STEM: Tall, smooth, sometimes waxy

NOTE: Native Americans use the plant ot treat pain and swelling.

Jewelweed

Family: Balsam (Balsminaceae)

Scientific Name: *Impatiens capensis* Height: 2–5' (0.6-1.5m)

FLOWER: Irregular, small, orange with rusty spots and large, flared open mouth that leads to a sac–like tube and spur; spur curves backward, the tube is half its length. Flowers are few and dangle loosely from arching overhead stalks.

LEAF: Simple, alternate, stalked, oval and tapered to tip, leaves thin, bright green on top, whitish beneath with coarse, rounded, forward–pointing teeth.

STEM: Slender, translucent green, succulent (water–filled)

NOTE: Annual; needs to re–seed each year. Ripe seed pods have mechanisms to shoot seeds at a touch. Self-pollinating. Juice on stem used by Native Americans for nettle stings and poison ivy.

Blue Vervain Family: Verbena (Verbenaceae)

Scientific Name: *Verbena hastata* Height: 2-6' (61-183 cm)

FLOWER: Regular, tiny, blue to purple, fused tube is trumpet–shaped with 5 spreading lobes, borne on numerous slender spikes arranged in candelabra style at tip and ends of angled side branches. Most abundant number of spikes at tip of main stem

LEAF: Simple, opposite, short stalked, lance–shaped to narrowly oblong coming to sharp tip; edges coarsely forward-pointing and sharp-toothed. Upper surface with stiff, short hairs. Lowest leaves may have 2 lobes at base.

STEM: Square–4 sided, grooved

NOTE: Native Americans used vervain as a diuretic and to treat headaches, circulatory issues and insomnia.

Brook Lobelia

Family: Bellflower (Campanulaceae)

Scientific Name: *Lobelia kalmii*

Height: 6–18" (15cm-46 cm)

FLOWER: Regular, tiny (⅓"-½"), blue with white centers, long stalked, flowers in loose racemes

LEAF: Simple, opposite, stalkless, narrow/entire/indistinctly toothed, upper leaves linear/grass-like, basal leaves broadly round then tapering toward the base (spatulate), may have small bracts at the base of each leaf

STEM: Tall and thin, usually branched, stiffly erect

NOTE: Poisonous. Also known as Kalm's Lobelia.

Great Blue Lobelia

Family: Bellflower (Campanulaceae)

Scientific Name: *Lobelia sipheletica* Height: 1–3' (30–91cm)

FLOWER: Irregular, medium, dark blue, fused tube split into two lips; upper lip has 2 erect lobes (ears), lower lip has 3 spreading lobes with blue and white stripes under the throat. Stalked flowers are held pointing upward in the leaf axils of the spike–like cluster.

LEAF: Simple, alternate, lance shape, stalkless, edges irregularly toothed, leaf size diminishes up the spike

STEM: Stout, erect, unbranched; smooth or few hairs

NOTE: Used in herbal medicine to treat respiratory problems, allergies, asthma, and bronchitis.

Northern Green Bog Orchid

Family: Orchid (Orchidaceae)

Scientific Name: *Platanthera aquilonis* Height: 2-23" (5-60 cm)

Status: protected

FLOWER: Small, 20-45, yellowish green, tightly packed along stem; sepals light green, arched over small green column; petals light green, curl up around column; lip small, flat

LEAF: 2-7 spreading narrow leaves up to 9" (23 cm) long

STEM: Strong, tall, single green stem

NOTE: Self-pollinating

LATE SUMMER WALK CHECKLIST

- O Bellflower, marsh
- O Boneset
- O Cardinal flower
- O Goldenrod, bog
- O Goldenrod, rough-leaved
- O Goldenrod, smooth
- O Grass of Parnassus
- O Jewelweed
- O Lobelia, brook
- O Lobelia, great blue
- O Nodding ladies' tresses
- O Northern green bog orchid
- O Purple-fringed orchid
- O Purple-stem aster
- O Round-leaved sundew
- O Spotted Joe-Pye weed
- O Swamp thistle
- O Tall white bog orchid
- O Turtlehead
- O Vervain, blue

FIELD NOTES

FIELD NOTES

FIELD NOTES

CHAPTER FOUR
EXPLORING WETLAND HABITATS

DELVE DEEPER INTO HABITATS

Now that you've spent some time enjoying wetland walks through the spring and summer, you may want to delve deeper and get better acquainted with four common wetland habitats.

First, you need to know where you are. Look around; what do you see? Are you standing in a cool, dark place surrounded by cedar trees and hummocks of sphagnum moss in the CEDAR SWAMP FOREST? Or winding your way through a tangle of shrubs in the NORTHERN SHRUB THICKET? Maybe you're out in the sun meandering through tall plants, some with flowers, some without, in the NORTHERN WET MEADOW. Or paddling along the shore of a lake or stream near the EMERGENT MARSH.

Each wetland habitat becomes wetter as you walk downhill toward the lake or stream. Get to know the plants and birds that are commonly found in each one and you'll become a wetland lover for life!

CEDAR SWAMP FOREST WALK

The **CEDAR SWAMP FOREST** is mostly shaded by evergreen branches. Feel the closed-in tightness of the trees. When the wind blows down trees, they create openings- favorite habitats for special plants that eat insects, such as the northern pitcher plant. In spring, listen for the "Feebee! Feebee! Feebee!" whistle of the black-capped chickadee and the drumming of the ruffed grouse attracting his mate. You won't see much water, but you'll smell it and the moist, earthy sphagnum moss. The cool, damp air may cause you to add another layer.

Sphagnum mosses are the dominant plants in several wetland habitats such as bogs and cedar swamp forests. Most of these mosses grow in large, compact cushions or wide, often saturated lawns just above the water table. A sphagnum plant can hold water equal to fifteen to twenty times its own weight. A given hummock may contain three or four different species of sphagnum moss. When sphagnum moss is harvested, it is dried and sold as peat moss. Gardeners use peat moss to amend the soil because it holds moisture.

As you peer into the shadows, don't you wonder how a white-tailed deer can bound through these jumbled tree limbs? Deer are so big, and the spaces are so small. You may see a spot where deer matted down the plants, perhaps last night. See the lush, green sphagnum moss on the forest floor. This plant cover may be hundreds of feet thick, and may have been growing here for thousands of years—amazing!

The Cedar Swamp Forest has trees that belong to two groups: deciduous and coniferous. Deciduous trees grow leaves in the spring and drop them in the fall. This group includes black ash, yellow birch, and red maple. Coniferous trees grow needles, bear their seeds in cones and usually stay evergreen year-round. This group includes northern white cedar, balsam fir, black spruce and tamarack. Tamarack is an exception to the rule. It grows needles and bears cones, but it is not evergreen. Instead, its needles turn golden in the fall and drop off the tree.

CEDAR SWAMP FOREST PLANTS

TREES

Deciduous Trees

Black ash
Red maple
Yellow birch

Coniferous Trees

Balsam fir
Black spruce
Northern white cedar
Tamarack

SPRING (April–May)

Shrubs

Sweet gale – white

Ferns & Fern Allies

Cinnamon fern
Interrupted fern
Royal fern

Flowers

Jack–in–the–pulpit – purple
Trailing arbutus – white

EARLY SUMMER (June–July)

Shrubs

Small cranberry – pink
Dwarf raspberry – white
Dwarf scouring rush – green

Ferns & Fern Allies

Lady fern
Oak fern

Flowers

Bunchberry – white
Calypso orchid – white
Canada mayflower – white
Creeping snowberry – white
Nude mitrewort – white
Poison ivy – white
Starflower – white
Grass pink – pink
Fringed polygala – pink
Pitcher plant – red
Pink lady's slipper – pink
Twinflower – pink
Purple meadow rue – purple
Northern green bog orchid – green
Clintonia – yellow
Goldthread – yellow
Large yellow lady's slipper – yellow
Small yellow lady's slipper – yellow
Northern swamp buttercup – yellow

LATE SUMMER (August– September)

Shrubs

Swamp rose – pink

Ferns & Fern Allies

Bulblet fern
Ostrich fern
Sensitive fern

Flowers

Cardinal flower – red
Swamp thistle – purple/pink

CEDAR SWAMP FOREST CHECKLIST

COMMON PLANTS

- O Balsam fir
- O Black ash
- O Black spruce
- O Bulblet fern
- O Bunchberry– white
- O Calypso orchid– white
- O Canada mayflower– white
- O Cardinal flower– red
- O Cinnamon Fern
- O Clintonia– yellow
- O Creeping snowberry– white
- O Dwarf raspberry– white
- O Dwarf scouring rush– green
- O Fringed polygala– pink
- O Goldthread– yellow
- O Grass pink– pink
- O Interrupted Fern
- O Jack–in–the–pulpit– purple
- O Lady Fern
- O Large yellow lady's slipper– yellow
- O Northern green bog orchid– green
- O Northern swamp buttercup– yellow
- O Northern white cedar
- O Nude mitrewort– white
- O Oak Fern
- O Ostrich fern
- O Pink lady's slipper– pink
- O Pitcher plant– red
- O Poison ivy– white
- O Purple meadow rue– purple
- O Red maple
- O Royal Fern
- O Sensitive fern

- O Small cranberry– pink
- O Small yellow lady's slipper– yellow
- O Starflower– white
- O Swamp rose– pink
- O Swamp thistle– purple/pink
- O Sweet gale– white
- O Tamarack
- O Trailing arbutus– white
- O Twinflower– pink
- O Yellow birch

COMMON BIRDS

- O Eagle, Bald (SC)
- O Flycatcher, Yellow-bellied
- O Grouse, Ruffed
- O Hawk, Cooper's
- O Hawk, Red-shouldered (T)
- O Hawk, Sharp-shinned
- O Heron, Green
- O Kingfisher, Belted
- O Kinglet, Golden-crowned
- O Kinglet, Ruby-crowned
- O Osprey
- O Owl, Long-eared (T)
- O Sparrow, White-throated
- O Swallow, Tree
- O Warbler, Blackpoll
- O Water thrush, Northern
- O Waxwing, Cedar
- O Woodcock, American

SC= species of concern

T= threatened

FIELD NOTES

FIELD NOTES

NORTHERN SHRUB THICKET WALK

The **NORTHERN SHRUB THICKET** is found typically between the cedar swamp forest and the wet meadow. But you can also find this plant community in other locations: drainage areas and roadside ditches, as well as along old railroad grades.

The tag/speckled alder dominates the thicket, but many other shrubs, flowering plants and ferns also grow here. Many of these species grow in the northern wet meadow and the cedar swamp forest as well. The graceful branches of tag alder along with red twig dogwood, pussy willow, winterberry, meadowsweet and high bush cranberry can be so dense that they're hard to walk through without a boardwalk.

Sensitive ferns, cinnamon ferns, royal ferns and marsh ferns are seen frequently in the thicket. Flowers such as marsh bellflower, marsh marigold, Joe-Pye weed, common boneset, jewelweed, wild blue flag, purple-stem aster, golden ragwort and purple meadow-rue are common in the shrub thicket.

These plants provide hiding places for many wintering small mammals and elusive birds. Listen for the "Poor Sam Peabody! Peabody! Peabody!" calls of the white-throated sparrow. Learning to identify the songs of the many wetland birds living in the thicket is a life-long challenge. A gentle breeze may bring the arresting aroma of the cedar, warmed by the heat of the sun. Often you can see pools of water - a clue you're getting closer to a pond, stream or lake. Most of the shrubs in the shrub thicket bloom in the spring.

NORTHERN SHRUB THICKET PLANTS

SPRING (April–May)

Shrubs

Pussy willow – white
Sweet gale – brown
Tag/Speckled alder – brown

Ferns & Fern Allies

Cinnamon fern
Interrupted fern
Royal fern

EARLY SUMMER (June–July)

Shrubs

Bog rosemary – white/pink
Buttonbush – white
Labrador tea – white
Leatherleaf – white
Meadowsweet – white
Red twig dogwood – white
Winterberry/Michigan holly – white
Poison sumac – yellow

Ferns & Fern Allies

Lady fern
Marsh fern
Oak fern

Flowers

Purple meadow rue – purple
Golden ragwort – yellow

LATE SUMMER (August–September)

Shrubs

Common elder/elderberry – white

Ferns & Fern Allies

Bulblet fern
Ostrich fern
Sensitive fern

Flowers

Blue flag iris – blue
Common boneset – white
Purple-stem aster – purple
Rough-leaved goldenrod – yellow
Spotted Joe-Pye weed – purple

NORTHERN SHRUB THICKET CHECKLIST

COMMON PLANTS

- O Blue flag iris – blue
- O Bog rosemary – white/pink
- O Bulblet fern
- O Buttonbush – white
- O Cinnamon fern
- O Common boneset – white
- O Common elder/elderberry – white
- O Golden ragwort – yellow
- O Interrupted fern
- O Labrador tea – white
- O Lady fern
- O Leatherleaf – white
- O Marsh fern
- O Meadowsweet – white
- O Oak fern
- O Ostrich fern
- O Poison sumac – yellow
- O Purple meadow rue – purple
- O Purple-stem aster – purple
- O Pussy willow – white
- O Red–twig dogwood – white
- O Rough-leaved goldenrod – yellow
- O Royal fern
- O Sensitive fern
- O Spotted Joe-Pye weed – purple
- O Sweet gale – brown
- O Tag/Speckled alder – brown
- O Winterberry/Michigan holly – white

COMMON BIRDS

O American Crow
O Blackbird, Red-winged
O Flycatcher, Great-crested
O Goldfinch, American
O Grosbeak, Evening
O Heron, Green
O Kingbird, Eastern
O Kingfisher, Belted
O Nuthatch, Red-breasted
O Owl, Great gray
O Phoebe, Eastern
O Sparrow, Swamp
O Swallow, Tree
O Warbler, Chestnut-sided
O Warbler, Northern parula
O Warbler, Yellow
O Waxwing, Cedar
O Woodpecker, Black-backed

FIELD NOTES

FIELD NOTES

NORTHERN WET MEADOW WALK

Coming out of the thicket, you may need to take off your shirt and put on your sunglasses as you enter the **NORTHERN WET MEADOW**. Dragonflies dart from place to place, feasting on the insects that feed on the tall, showy wildflowers waving in the breeze between the shrub thicket and the water's edge. The wet meadow is covered with grasses and grass-like plants called sedges. Enjoy the fragrant shrubs: swamp rose, sweet gale, and bog rosemary. Listen for the "Sweet! Sweet! Sweet! I'm so sweet!" call of the yellow warbler.

The northern wet meadow has shallow, standing water year-round and contains two general groups of flowering plants: showy flowers and inconspicuous flowers.

Inconspicuous flowers are found on rushes, reeds, grasses and sedges. Because these plants are easily confused, they are difficult to identify. I've introduced you to them, so you can begin to appreciate their diversity and importance in wetlands.

The northern wet meadow is dominated by sedges, so it is sometimes called a sedge meadow. There are more than 200 sedge species in the Great Lakes region.

CAUTION: Be careful touching wet meadow plants. Sedges have edges, so sharp they'll cut you!

Here are a couple of ways to help you distinguish sedges, grasses and rushes:

> *"Sedges have edges; rushes are round.*
>
> *Grasses are hollow; what have you found?"*

The northern wet meadow also has wonderful, showy, tall wetland wildflowers. Many of these plants have relatives that grow in the dry meadow further upland - like thistles and milkweeds, while others have relatives in Asia and Europe. Some of these distant relatives have found their way into our native plant communities and are now considered invasive species. Most northern wet meadow flowers bloom in late summer.

NORTHERN WET MEADOW PLANTS

SPRING (April–May)

Shrubs

Shrubby cinquefoil – yellow

Ferns & Fern Allies

Cinnamon fern
Royal fern

EARLY SUMMER (June–July)

Flowers

Wood lily – orange
Dragon's mouth orchid – pink
Grass pink – pink
Showy lady's slipper – pink
Swamp milkweed – pink
Marsh thistle/European swamp thistle **(invasive)** – purple
Purple avens – purple

LATE SUMMER (August– September)

Ferns & Fern Allies

Sensitive fern

Sedges

Porcupine sedge
Tussock sedge

Flowers

Boneset – white
Grass of Parnassus – white
Rough bedstraw – white
Tall white bog orchid – white
White turtlehead – white

Jewelweed – orange
Bog goldenrod – yellow
Smooth goldenrod – yellow

Spotted Joe–Pye weed – pink
Great blue lobelia – blue
Small purple–fringed orchid – purple
Swamp thistle – purple

NORTHERN WET MEADOW CHECKLIST

COMMON PLANTS

- O Bog goldenrod – yellow
- O Boneset – white
- O Cinnamon fern
- O Dragon's mouth orchid – pink
- O Grass of Parnassus – white
- O Grass pink – pink
- O Great blue lobelia – blue
- O Jewelweed – orange
- O Marsh thistle/ European swamp thistle **(invasive)** – purple
- O Porcupine sedge
- O Purple avens – purple
- O Riddell's goldenrod – yellow
- O Rough bedstraw – white
- O Royal fern
- O Sensitive fern
- O Showy lady's slipper – pink
- O Shrubby cinquefoil – yellow
- O Small purple–fringed orchid – purple
- O Smooth goldenrod – yellow
- O Spotted Joe–Pye weed – pink
- O Swamp milkweed – pink
- O Swamp thistle – purple
- O Tall white bog orchid – white
- O Tussock sedge
- O White turtlehead – white
- O Wood lily – orange

COMMON BIRDS

- O Bittern, American
- O Blackbird, Red-winged
- O Crane, Sandhill
- O Flycatcher, Alder
- O Heron, Great blue
- O Hawk, Marsh/Northern Harrier (SC)
- O Owl, Short-eared (E)
- O Rail, Virginia
- O Sandpiper, Spotted
- O Snipe, Wilson's/Common
- O Sora
- O Sparrow, Swamp
- O Swallow, Bank
- O Warbler, Canada
- O Warbler, Mourning
- O Warbler, Common yellowthroat
- O Wren, Marsh (SC)
- O Wren, Sedge

SC= species of concern
T= threatened

FIELD NOTES

FIELD NOTES

EMERGENT MARSH

Finally, you can see the open water. Along the shore, at the bay ends of lakes and in the bends of rivers and streams, tough-rooted **EMERGENT MARSH** plants grow. Listen to the winds and lapping waves of the moving water. These waves and currents push and pull these flexible, soft-stemmed plants, but their roots hold firm in the mucky, water-logged soil. Red-winged blackbirds perch on top of cattails and bulrushes and sing, "konk-la-ree," across the marsh. Great blue herons wade through reeds and rushes at the water's edge. Wild rice and cotton grass hold their ground throughout the year but die back in winter. Used by muskrats, mink, otters and beavers, as well as fish, the emergent marsh provides food and shelter for many wildlife species.

Marshes have shallower, slower moving water than the deeper, faster moving open water of the river or lake. The water depth varies from several feet deep to less than an inch. The water level in a marsh can fluctuate seasonally or from year to year.

Marshes are an important transition zone between lake and shore. Plants growing in open water are called aquatic plants. As these aquatic plants die, they add material to the bottom sediments. This accumulation slows the water, causes depth to decrease and slows down water movement. These new conditions are a perfect environment for marsh plants.

The best way to see aquatic plants and marsh plants is by boat. Look for muskrat mounds, beaver lodges and turtles basking in the sun when you take a cruise.

Because boats often carry plant seeds or plant parts into the water, marshes are one of the first places where invasive species become established. I've paired common native and non-native species together, so you can begin to tell them apart.

I've included some aquatic plants and two groups of marsh plants: those with inconspicuous flowers and showy flowers.

EMERGENT MARSH PLANTS

Plants are organized into three groups: showy flowering marsh plants, inconspicuous flowering marsh plants and aquatic plants. They are grouped by season and listed by color, so you can find them easily.

SHOWY FLOWERING MARSH PLANTS have brightly colored flowers that are a pleasure to see.

SPRING (April–May)

Flowers

Marsh marigold – yellow

EARLY SUMMER (June–July)

Ferns & Fern Allies

Marsh fern

Flowers

Marsh bellflower – white

Marsh cinquefoil – yellow
Yellow flag iris **(Invasive) –** yellow
Yellow loosestrife/ Swamp candles – yellow

Blue flag iris – blue
Marsh thistle/European swamp thistle **(Invasive)** – purple
Purple loosestrife **(Invasive)** – purple

LATE SUMMER (August–September)

Flowers

Bulb-bearing water-hemlock– white
Swamp thistle– purple

FIELD NOTES

INCONSPICUOUS FLOWERING MARSH PLANTS have green or brown flowering parts. These plants provide important food for birds and insects, as well as habitat for fish, frogs and turtles. Wetland mammals such as muskrats, otters and beavers collect building materials from the marsh.

EARLY SUMMER (June–July)

Bur-reed
Soft-stem bulrush
Broad-leaved cattail
Narrow-leaved cattail **(Invasive)**
Common reed/native phragmites

LATE SUMMER (August–September)

Reed canary grass/non-native phragmites **(Invasive)**

INCONSPICUOUS FLOWERING MARSH PLANTS

Bur-reed

Broad-leaved cattail

Native phragmites

Bulrush

Narrow-leaved cattail

Invasive phragmites

AQUATIC PLANTS can be organized into three groups, so depending on where you find them, they'll be easy to recognize.

- Free-floating plants
  Common duckweed

- Submergent plants growing just below the surface
  Common water–milfoil– red
  Eurasian water–milfoil– pinkish red
  Common water weed - white

- Emergent plants growing above the surface
  White water–lily– white
  Northern arrowhead– white
  Water smartweed– pink
  Bullhead water–lily– yellow
  Common bladderwort– yellow
  Pickerel weed– blue
  Water hyacinth **(Invasive)** – blue

EARLY SUMMER (June–July)

Common water weed– white
White water–lily– white
Water smartweed– pink
Bullhead water–lily– yellow
Pickerel weed– blue

LATE SUMMER (August–September)

Northern arrowhead– white
Common bladderwort– yellow
Water hyacinth **(Invasive)** – blue

AQUATIC PLANTS

Duckweed

Water-milfoil

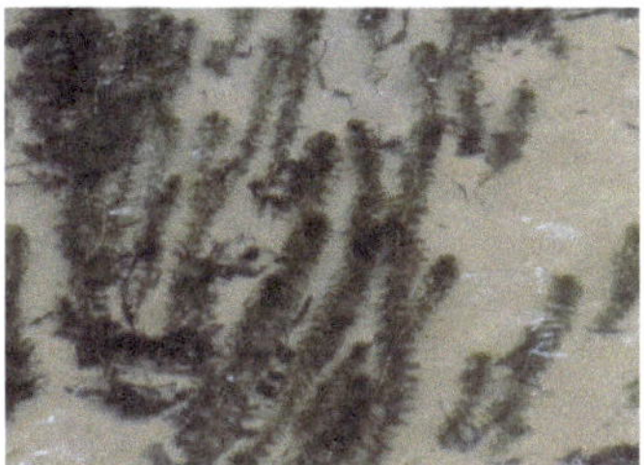
Water weed

White water-lily

Arrowhead

Smartweed

Bullhead water-lily

Bladderwort

Pickerel weed

Water hyacinth

EMERGENT MARSH CHECKLIST

COMMON PLANTS

- O Blue flag iris
- O Broad-leaved cattail
- O Bulb-bearing water-hemlock
- O Bullhead water-lily
- O Bur-reed
- O Common bladderwort
- O Common duckweed
- O Common water-milfoil
- O Common water weed
- O Common reed/native phragmites
- O Eurasian water-milfoil (**Invasive**)
- O Marsh bellflower
- O Marsh cinquefoil
- O Marsh fern
- O Marsh marigold
- O Marsh thistle/ European swamp thistle **(Invasive)**
- O Narrow-leaved cattail (**Invasive**)
- O Northern arrowhead
- O Pickerel weed
- O Purple loosestrife (**Invasive**)
- O Reed canary grass/non-native phragmites (**Invasive**)
- O Soft-stem bulrush
- O Swamp thistle
- O Water hyacinth (**Invasive**)
- O Water smartweed
- O White water-lily
- O Yellow flag iris (**Invasive**)
- O Yellow loosestrife/swamp candles

COMMON BIRDS

- O Bittern, American
- O Blackbird, Red-winged
- O Crane, Sandhill
- O Flycatcher, Alder
- O Heron, Great blue
- O Hawk, Marsh/Northern Harrier (SC)
- O Owl, Short-eared (E)
- O Rail, Virginia
- O Sandpiper, Spotted
- O Snipe, Wilson's/Common
- O Sora
- O Sparrow, Swamp
- O Swallow, Bank
- O Warbler, Canada
- O Warbler, Mourning
- O Warbler, Common yellowthroat
- O Wren, Marsh (SC)
- O Wren, Sedge

FIELD NOTES

FIELD NOTES

CHAPTER FIVE
SEARCHING FOR HIDDEN TREASURES

Deep in the forested wetlands, you'll find three rare treasures: the Canadian Carpet, Orchids and Insect-eating plants. So look closely and enjoy what you find!

CANADIAN CARPET

Sometimes when you're walking through the cedar swamp forest looking for showy wetland wildflowers, you may miss one of the hidden treasures in this wetland plant community. If you look closely, you may notice a dense green carpet beneath the cedar trees and find an intricate mosaic of mosses and small trailing plants growing over old stumps and decaying logs. Get down on your hands and knees or lie down flat on the boardwalk. Breathe in the cool, fragrant moistness of the Canadian Carpet!

The Canadian Carpet is a name given to a group of plants that share similar requirements and are adapted to living and thriving in close association with each other. These low-growing, compact flowering plants are very well adapted to life in the damp, shady swamp. The intertwining vines, trailing stems and widely-spreading roots are so interconnected you can't pull them apart. Most of the flowering plants are only a few inches high with long, tough underground stems. Their thick, waxy leaves protect and insulate them all winter under a thick, blanket of snow. They grow very slowly - a quarter of an inch is considered a good growing season.

The Canadian Carpet provides food for many animals living in the swamp year round. Mice, chipmunks and grouse eat the berries. Eastern cottontail rabbits and snowshoe hares love the leaves.

The range of the Canadian Carpet is extremely limited. Today, the Canadian Carpet is found only in the southern half of Canada and in the northern United States, as well as in parts of southern Scandinavia and northern Russia. We are lucky to have it in the northern half of the Great Lakes basin, above the 45th parallel.

FINDING THE CANADIAN CARPET

The best way to find the Canadian Carpet is to know which plants grow there. Look up the plants in the Canadian Carpet Checklist and become familiar with them. This will help you recognize some of these plants when you're walking through the swamp. Then, notice which other plants are growing with them. It's like a true treasure hunt!

Whichever combination of Carpet plants you find, you'll see this characteristic intertwining and close association, which is their signature.

APPRECIATING THE CANADIAN CARPET

Virginia Eifert described the Canadian Carpet in her book _Journey in Green Places_. Her vivid writing helps us all see the value in something we would miss otherwise.

> "The very carpet is the signature of the place, as if seeing it I did not need the presence of spruce or of fir… or the lingering breath of the last glacier, to tell me what to expect here." (Eifert, page 51)
>
> "It was a logical thing to find-always in its proper place, because it simply will not grow anywhere else, identifying that tremendous upholstering which was called forth by the post-glacial replanting following along in the damp footsteps of the great ice." (Eifert, page 46)
>
> "The Carpet is intricately woven. It usually carries part of its pattern as an identifying mark, yet every woods has a somewhat different design, some variation, like true Oriental rugs, of which no two are ever created quite alike." (Eifert, page 50)

CANADIAN CARPET CHECKLIST

FIELD NOTES

FIELD NOTES

ORCHIDS

Before you leave the shade of the cedar swamp forest, look for another hidden treasure - orchids. As you emerge from the swamp and shrub thicket, you may find yourself in an opening with sun-loving orchids.

You may need to take off a layer and put on some sunscreen or consider taking this walk in the morning when it's cooler and less buggy, too!

Orchids are the world's largest plant family with more than 30,000 species. Most people think of orchids growing in tropical places. Even though 80% of the orchids on Earth are native to the tropics, more than 50 species are found in the north, and the Great Lakes basin has many of them. Orchids in northern wetlands are easy to identify by their pouched or lipped flowers and leaves with parallel veins.

Orchids are fragile and rare!

- **Each kind of orchid is pollinated by only 1 or 2 kinds of insects.** When insects such as bees, moths and butterflies disappear because of pesticides, so do the orchids.

- **Orchids have the smallest seeds of any flowering plant and they need help from fungi.** Seeds are so small they look like specks of dust. One flower can produce millions of seeds. These tiny seeds are covered by a thin cell wall, which makes them easy to carry away on the wind and water. BUT, they lack stored food. With no self-contained food source, orchid seeds need help. They get help from fungi living in the soil. This relationship between orchids and fungi is so critical that it makes orchids difficult (if not impossible) to transplant from the wild.

- **Orchids are slow growing.** The seed grows underground for 3-7 years before it can produce above-ground leaves and a flower stalk. Once above ground, the plant may take another 5-10 years to develop a flower. When you see an orchid that is a foot tall, you are looking at a plant that has taken 10-15 years to reach this height! Orchids will not grow if they are injured by frost, trampling or disturbance.

- **Orchids are protected by law.** Do your part to help protect them:

    - Don't pick orchid flowers!
    - Don't dig up orchid plants!
    - Don't walk where orchids grow!

SUN LOVING ORCHIDS

May-July	Dragon's mouth orchid	Page 75
June-August	Grass pink	Page 77
	Rose pogonia	Page 79
	Showy lady's slipper	Page 80
July-August	Nodding ladies' tresses	Page 109
	Purple-fringed orchid	Page 116
	Tall white bog orchid	Page 110

SHADE LOVING ORCHIDS

May-June	Calypso orchid	Page 74
	Pink lady's slipper	Page 78
	Small yellow lady's slipper	Page 90
May-July	Large yellow lady's slipper	Page 89
June-August	Northern green bog orchid	Page 125

ORCHIDS CHECKLIST

- O Calypso orchid - shade
- O Dragon's mouth orchid - sun
- O Grass pink - sun
- O Nodding ladies' tresses - sun
- O Northern green bog orchid - shade
- O Pink lady's slipper - shade
- O Purple-fringed orchid - sun
- O Rose pogonia - sun
- O Showy lady's slipper - sun
- O Tall white bog orchid - sun
- O Yellow lady's slipper, large - sun/shade
- O Yellow lady's slipper, small - sun/shade

FIELD NOTES

FIELD NOTES

INSECT EATERS

Insect-eating plants may be the most difficult hidden treasure to find on your wetland walk through the cedar swamp forest. These plants live in sunny, open, wet places with acidic water that lacks the nutrients they need, so they compensate by capturing and digesting insects for food.

A plant can eat an insect in four simple steps:

1. Attract insects such as mosquitoes, gnats and midges.
2. Capture the insects.
3. Prevent the insects from escaping.
4. Digest the insect's body and use it as food.

Here is how these hidden treasures do their magic!

The pitcher plant attracts insects in two ways: First, with its scented nectar, and second, with the water it collects in the bottom of the pitcher-shaped leaf. When the insect lands on the spout-shaped edge of the leaf, it slides down the slick leaf walls and into the water. Downward-facing hairs on the inside of the leaf walls prevent the insect from crawling out of the leaf. Chemicals in the water at the bottom of the leaf break down the insect, so the plant can absorb the nutrients it needs to survive.

The sundew attracts insects with a "deadly dew" secreted from the gland-tipped hairs sticking up on the leaves. This sticky, gel-like substance traps and holds the insects. As the bug struggles to free itself, the sundew senses the movement of the hairs. The edges of the leaf slowly wrap around the insect and form a temporary "stomach." As it digests the insect, the plant gains the vital nitrogen it needs to grow.

Look for these insect-eaters growing in sphagnum moss among black spruce and horsetails.

Northern Pitcher plant Page 73
Round-leaved sundew Page 113

Pitcher Plant's pitcher-shaped leaf

INSECT EATERS CHECKLIST

- O Northern pitcher plant- sun
- O Round-leaved sundew- sun

Sundew's gland-tipped hairs on leaves and its tiny white flower

FIELD NOTES

CHAPTER SIX
FERNS & FERN ALLIES

FERN WALK

If it's not a wildflower, a shrub or a tree, what else could it be? Maybe it's a fern. You may already be familiar with bracken ferns growing in open, dry fields or delicate maiden hair ferns growing in the beech-maple forest.

Ferns are ancient plants. Hundreds of millions of years ago, ferns and fern allies provided nutritious food for plant-eating dinosaurs that roamed the Earth. Some ferns grew as big as trees. Over thousands of years, these swamp-fern forests died and piled up layer upon layer, miles thick. The layers eventually compressed and turned the plants into coal.

Ferns were growing in the Great Lakes Watershed long before the conifer forests we know today. Some ferns have been in the swamp since the glaciers left more than 10,000 years ago. Their deep, underground, root-stocks are called rhizomes.

Rhizomes produce roots below the ground and upright stems above the ground called fronds. Each spring new fronds grow out of the rhizomes. These tightly curled fronds are called "fiddle heads," because they are shaped like the end of a fiddle. Fiddle heads were gathered by Native Americans and eaten in the early spring.

Because ferns are such ancient plants, they've had millions of years to develop unique structures and patterns that are incredibly complicated. The study of ferns is organized in three ways:

1) Fern Anatomy
2) Clues for looking at ferns
3) Easily identified wetland ferns

FERN ANATOMY

Fern Parts

Ferns have plant parts that may not be familiar to you, but they're worth knowing.

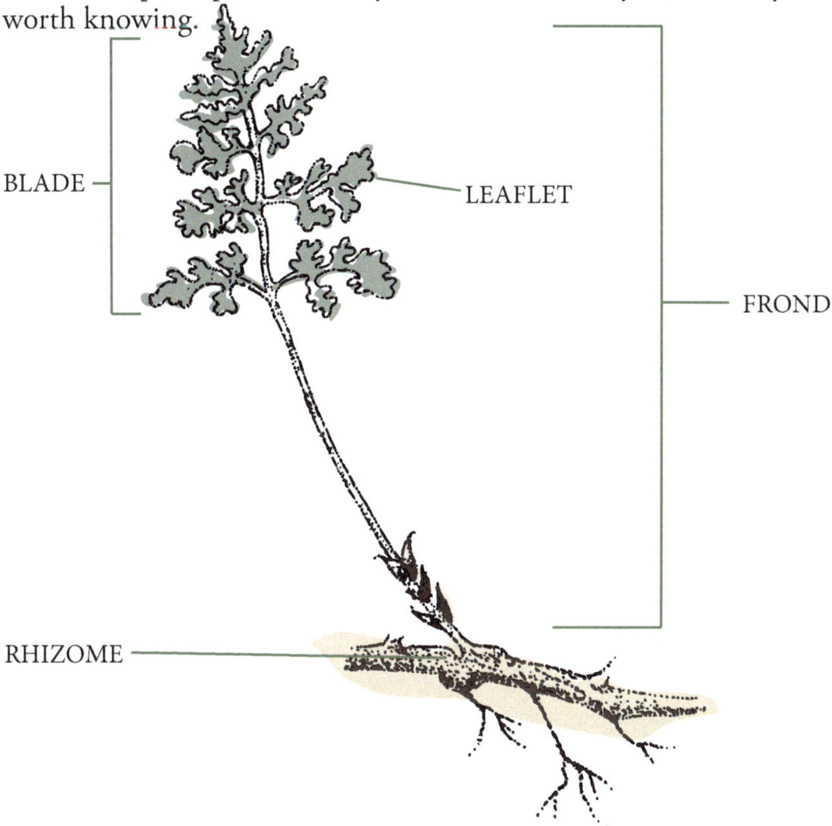

FROND-the whole fern leaf, including both plant stem and blade

BLADE- the entire leafy part of the frond

LEAFLET- a separate division of the frond that looks like small leaves, each leaflet is attached to a stem

RHIZOME- the underground root stalk

WINGED STEM- a plant stem with leafy tissue that sticks out and looks like a wing

Fern Reproductive Parts

Ferns reproduce with spores. Spores are special reproductive cells that can be located on different parts of a fern.

Most often spores are located on the underside of a particular frond, called a fertile frond. All of the leaflets on a fern frond may not have spores, but if any leaflets have spores, then it is a fertile frond.

Turn the fertile frond over and look at the leaflets. You may discover small round dots that turn brown as they age. These dots are really fruit dots containing spores. When you look more closely with a hand lens, you'll see an amazing variety of shapes and patterns.

Sometimes spores are bunched together on a separate stalk, called a fertile stalk. You may or may not see a fertile stalk depending on the time of year and the type of fern.

Fertile stalk (sensitive fern)

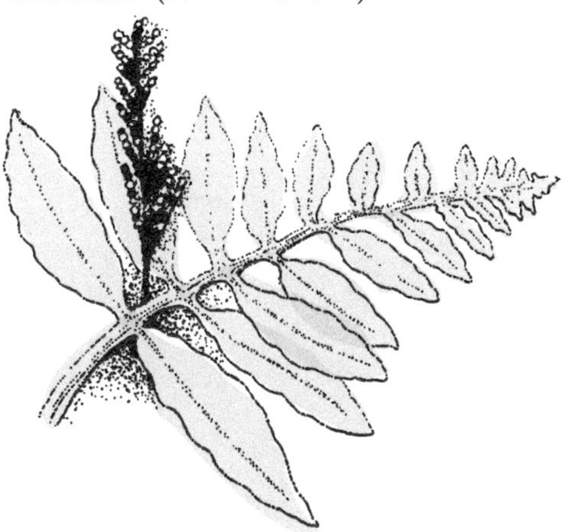

Fertile frond (marsh fern)

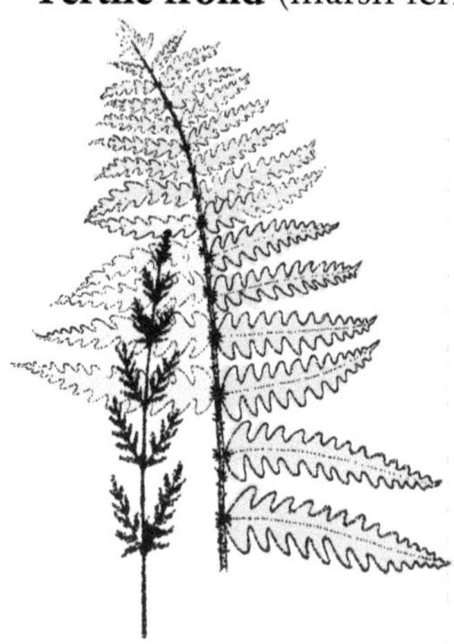

Fern Frond Shapes

Fern fronds have different shapes. These shapes are helpful in remembering different kinds of ferns.

Broadest at base

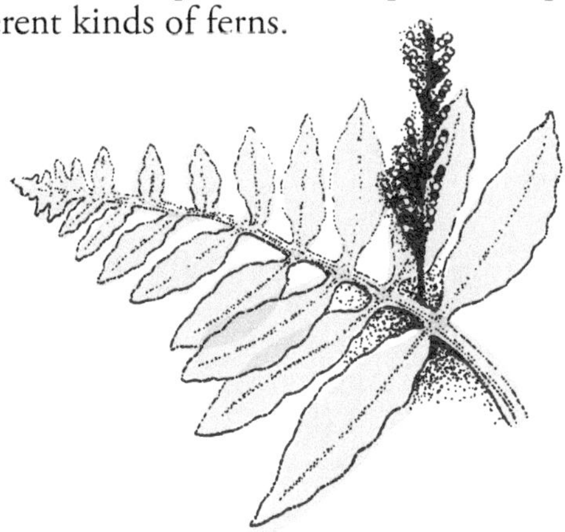

Tapered all of the way to the base

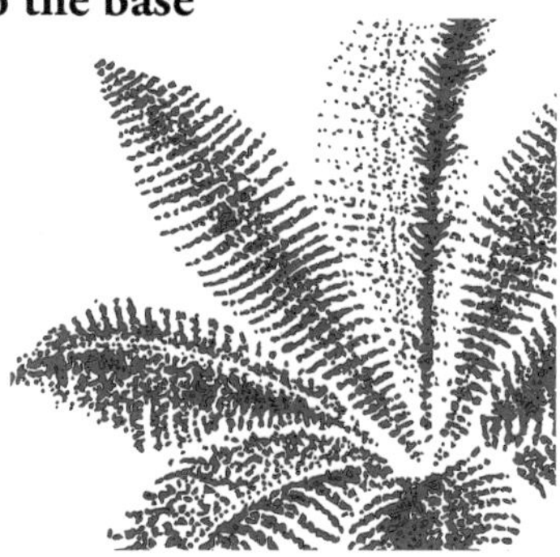

Tapered part of the way to the base

FERN CLUES

Hunting for Ferns in Wetlands

When you choose a fern to study, make sure it is a normal, mature specimen. If a number of ferns are growing in a group, choose a large, healthy-looking one. Then, look for the following clues:

Clue #1- Habitat

You can find many species of ferns in wetlands. Some ferns like their "feet" out of the water, while others prefer to grow in standing water.

Ferns also prefer different amounts of sunlight. Some ferns like to grow in the low or filtered light of the cedar swamp forest or shrub thicket, while others prefer the full sun of the wet meadows and marshes.

Shade- Loving Ferns prefer their "feet" out of water. They grow in the cool dark cedar swamp forest or shrub thicket, as well as along streambanks.

Bulblet fern
Interrupted fern
Lady fern
Ostrich fern
Sensitive fern
Oak fern

Sun- Loving Ferns prefer their roots standing in water. They usually grow in open, light marshes and wet meadows, as well as along lakeshores.

Cinnamon fern
Marsh fern
Royal fern

Clue #2- Growth Habit

Look at how the fern is growing. Are the fronds standing all alone and scattered, or growing together in a clump?

Clue #2- Blade Shape

Is the frond broadest at the base, tapered all of the way to the base or tapered part way down the stalk?

Clue #3- Fertile Part

Is there a fertile stalk or a fertile frond?

Clue #4- Plant Height

Is the fern low-growing (L), medium height (M) or tall (T)?

Low-growing Ferns: Oak fern, Bulblet fern

Medium Height Ferns: Lady fern, Marsh fern, Sensitive fern

Very Tall Ferns: Ostrich fern, Cinnamon fern, Royal fern, Interrupted fern

Bulblet fern Habitat: Shade or partial shade

Scientific Name: *Cystopteris bulbifera* Height: 1.0-3.5', 30-105cm

GROWTH HABIT: clumped but not vase-shaped, low-lying, lax fronds spread in all directions from the center of the clump

BLADE SHAPE: leaflets don't grow all the way to the base of the frond. The frond tapers toward the elongated tip.

FERTILE PART: This fern is unique because it is the only species in the northeastern U.S. that reproduces in two different ways: a fertile frond with few fruit dots and small bulblets underneath the blade that drop off and germinate into new plants.

DISTINGUISHING CHARACTERISTICS:
#1- Long, narrow, thin leaves like streamers trail along the ground.
#2- Bulblets grow on the underside of the blade.

Cinnamon fern Habitat: shade or partial shade

Scientific Name: *Osmunda cinnamomea* Height: 2-5', 60-150cm

GROWTH HABIT: clumped and vase-like with arching tips, tall

BLADE SHAPE: tapered part way to the base

FERTILE PART: Fertile stalks grow from the center of the clump.

DISTINGUISHING CHARACTERISTICS:
#1- Woolly tufts grow where leaflets join the stem.
#2- Fertile stalks first to appear bright green in spring, turning to lustrous cinnamon brown and withering in early summer.
#3- Base of the stem is covered with cinnamon brown wool. Hummingbirds use this wool to line their nests.
#4- Bottom-most leaflets angle downward.

Interrupted fern

Habitat: shade or partial shade

Scientific name: *Osmunda claytoniana*

Height: 2-4', 60-120cm

GROWTH HABIT: vase-shaped clump, tall

BLADE SHAPE: tapered part way to the base

FERTILE PART: The fertile frond is woolly brown in the early spring as the fiddle heads are unfurling. The fertile parts interrupt the blade. When the fertile parts ripen and wither, they leave a space on the blade from early summer onward.

DISTINGUISHING CHARACTERISTICS:
#1- Fertile leaflets leave an interrupted space on the blade.
#2- Fertile frond is much longer than sterile frond & more erect.
#3- Bottom-most leaflets angle upward.

Lady fern

Habitat: partial shade

Scientific name: *Athyrium filix-femina*

Height: 1.5-3.5', 45-105cm

GROWTH HABIT: vase-shaped clump, medium size

BLADE SHAPE: tapered part way to the base

FERTILE PART: Fertile frond has long, angled fruit dots on the leaflets.

DISTINGUISHING CHARACTERISTICS:
#1- Fronds are finely cut and very lacy.
#2- The pointed tips droop downward.
#3- Fruit dots on the fertile frond form a herringbone pattern.

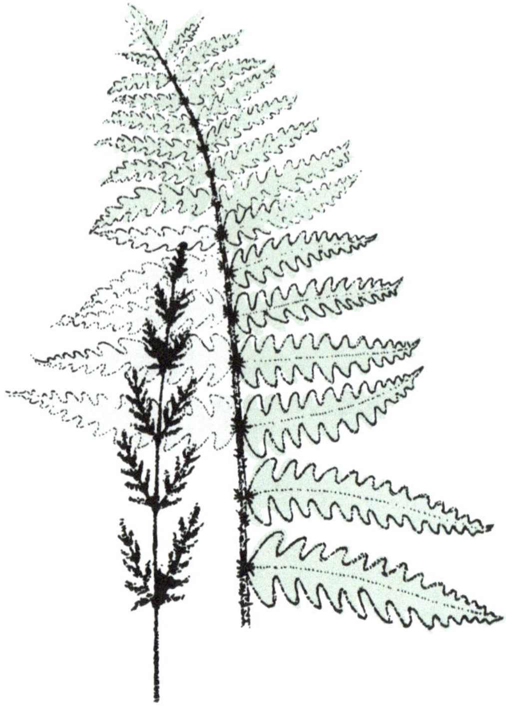

Marsh fern Habitat Full sun or partial shade

Scientific name: *Thelypteris palustris* Height: 1.0-2.5', 30-75cm

GROWTH HABIT: scattered, uncoiling new fronds grow up among fully developed fronds throughout the summer, medium size

BLADE SHAPE: tapered part way to base, widest almost at the base

FERTILE PART: Fertile fronds grow longer and more erect than sterile fronds. The edges of the leaflets curve over and cover the fruit dots.

DISTINGUISHING CHARACTERISTICS:
#1- Lowest pair of leaflets and other lower leaflets turn so they are almost perpendicular to stem, looking like a ladder or stairs.
#2- Thin, flexible stems enable the fronds to twist and turn in the breeze.

Oak fern Habitat: shade

Scientific name: *Gymnocarpium dryopteris* Height: 6-16", 15-40cm

GROWTH HABIT: scattered, often form a delicate low groundcover

BLADE SHAPE: broad-based triangle

FERTILE PART: The fertile frond has round fruit dots very near the edge of the leaflets. Fiddle heads are produced all summer.

DISTINGUISHING CHARACTERISTICS:
#1- The blade bends almost horizontally at its base.
#2- Each frond is branches into 3 distinct parts.
#3- The frond is yellow-green and looks translucent.

Ostrich fern Habitat: Full sun or partial shade

Scientific name: *Matteuccia struthiopteris* Height: 2-5', 60-150cm

GROWTH HABIT: vase-shaped clump, tall

BLADE SHAPE: tapered to the base, the base tapers to a point

FERTILE PART: Fertile fronds are smaller than sterile fronds. They are very stiff and appear to be woody. The fertile leaflets contract and fold over the fruit dots to form a series of closely packed, very hard beads that all together form a "pod." These pods are arranged on the fertile frond, so they appear to look like an ostrich feather.

DISTINGUISHING CHARACTERISTICS:
#1- Fertile frond resembles an ostrich feather.
#2- Sterile fronds surround the fertile fronds.

Royal fern Habitat: Sun, partial shade or full shade

Scientific name: *Osmunda regalis* Height: 2-5', 60-150cm

GROWTH HABIT: clumped and shrub-like, Tall (May grow over 6" in acid bogs.)

BLADE SHAPE: branching

FERTILE PART: The fertile frond has light brown, flower-like fertile leaflets at the top of the blade.
The fertile leaflets have densely clustered spore cases.

DISTINGUISHING CHARACTERISTICS:
#1- Fertile part is located at tip of blade.
#2- Fern frond is branching.
#3- Sterile leaflets are oblong and oval-shaped.
#4- Fertile leaflets are opposite and sterile leaflets are alternate.

Sensitive fern Habitat: Full sun or shade

Scientific name: *Onoclea sensibilis* Height: 1.0-2.5', 30-75cm

GROWTH HABIT: scattered, medium size

BLADE SHAPE: almost triangular but just below the widest part, the blade begins to taper partway to the base

FERTILE PART: The fruit dots are grouped in bead-like cases on the separate fertile stalk.

DISTINGUISHING CHARACTERISTICS:
#1- Leaflets are nearly opposite, broad and wavy.
#2- The stem has wings that get broader toward the tip.
#3- Lower pairs of leaflets are not winged.
#4- The fertile stalk is beaded and is present throughout the winter. Look for them when you're skiing or snowshoeing.

FERN CHECKLIST

- O Bulblet fern
- O Cinnamon fern
- O Interrupted fern
- O Lady fern
- O Marsh fern
- O Oak fern
- O Ostrich fern
- O Royal fern
- O Sensitive fern

FIELD NOTES

FIELD NOTES

LOOKING AT FERN ALLIES

Some plants are closely related to ferns, so we call them fern-allies. Fern-allies such as horsetails, clubmosses and mosses are sometimes mistaken for ferns. Fern-allies are smaller forms of much larger, ancient plants that lived 200-300 million years ago. Fossils of these huge plants can be seen in natural history museums.

Ferns and fern-allies covered much of the Earth long before there were flowering plants. Without flowers and seeds, they developed very special parts for reproduction. Like ferns, they produce spores - microscopic reproductive cells. When you look at a fern-ally, you will often notice a cone-like structure at the tip of a stalk that contains spores.

HORSETAILS

Horsetails are among the oldest kinds of plants on Earth and used to grow as large trees during the age of the dinosaurs. Some look like miniature bamboo, while others look like a pad for scouring dirty cooking pots.

A single horsetail plant can have two different kinds of ridged, rough stems. In the spring, a fertile stem with no leaves emerges. Spores form in a cone-like part at the tip of the stem. After the spores are released, the fertile stem shrivels. Later, a new sterile stem with leaves appears. These thin, wiry leaves form a whorl at the collar of each joint on the stem. Both kinds of stems are hollow with joints between each segment.

Water Horsetail (*Equisetum fluviatile*)
Dwarf Scouring Rush (*Equisetum scirpoides*)

CLUBMOSSES

Clubmosses are not true mosses, even though they may look like them. You may see what looks like many, small evergreen trees growing in a row but you're actually looking at a single plant.

The plant has runners growing over the surface of the ground with upright stems. Some stems may form branches with cone-like tips. The tips produce spores, which can grow into new plants.

Northern fir-moss (*Huperzia selago*)
Shining fir-moss (*Huperzia lucidula*)
Tree clubmoss (*Dendrolycopdium spp.*)

Water horsetail Family: Horsetail (Equisetaceae)

Scientific Name: *Equisetum fluviatile* Height: 3'+ (0.9m)

CONES: 1-2 cm long at tips of stems, long-stalked, blunt-tipped, deciduous

SHEATH: Green, 6-10 mm long, 12-24 teeth, dark brown to black, sometimes with narrow white margins

STEM: annual, fertile and sterile stems alike, 9-2 shallow, smooth ridges, large central cavity about 4/5 stem diameter, branches none or few, to many and regularly whorled from middle nodes, spreading without branchlets, hollow

NOTE: Descendants of ancient ferns, developed 270 million years ago. Some of the oldest plants on Earth.

Dwarf scouring rush

Family: Horsetail (Equisetaceae)

Scientific Name: *Equisteum scirpoides*

Height: 1-3" (3-7cm)

CONES: 1.5-3 cm long, stalked, blunt-tipped, deciduous

SHEATH: Green at base, red-brown and flaring at tip, teeth brown, joined in 3-5 broad

STEM: annual, erect, fertile and sterile stems not alike. Sterile stems green 3-7dm tall and 1.5-3 mm wide, with 10-18 ridges, rough to touch with sharp, hooked silica spines, central cavity ½-2/3 stem diameter. Fertile stems at first pink-brown, fleshy, unbranched, becoming green and branched.

NOTE: Smallest horsetail. Native people used it to clean pots.

Northern fir-moss Family: Clubmoss (Lycopodiaceae)

Scientific Name: *Huperzia selago* Height: 2-8" (6-20 cm)

SPORES: Similar to leaves, sporangia produced early in spring in leaf axils, followed by sterile leaves, upper axils produce small, 2-lobed reproductive buds

LEAF: Evergreen, yellow-green, in 8-10 rows, swollen and concave at base, gradually tapered to tip, mostly without teeth, hollow, appressed to stem, fertile leaves bear spore sacs, sterile leaves lack spores, may appear whorled

STEM: horizontal, short, upright, single or multiple from base, tips may curve up

NOTE: Easily adapts to a variety of habitats.

Shining fir-moss Family: Clubmoss (Lycopodiaceae)

Scientific Name: *Huperzia lucidula* Height: 10" (25 cm)

SPORES: With small teeth or entire at tip, buds in some upper leaf-axils

LEAF: Evergreen, mostly in 6 rows, spreading or curved downward, alternating groups of longer sterile leaves and shorter fertile leaves

STEM: evergreen, light green, creeping and rooting, upcurved stems forked several times, crowded with shiny, dark leaves

NOTE: Distinguished from Northern fir-moss by ragged look.

Tree Clubmoss

Family: Clubmoss (Lycopodiaceae)

Scientific Name: *Dendrolycopodium spp.* Height: 6-12" (15-30cm)

SPORES: Develop in spike-like or cone-like structures, ½"-2" long, stalkless at branch tips, 107 clustered at top of plant

LEAF: Evergreen, alternate, spirally arranged, ¼" long, medium green, toothless, prickly

STEM: Evergreen, horizontal, run underground, with horizontal branches

NOTE: Often called ground pine, because it looks like a small pine tree seedling.

FIELD NOTES

FIELD NOTES

MOSSES

Mosses are different from all the other plants in this book because they do not have roots, true stems or leaves with veins. Mosses have only two kinds of cells: one for making food and one for soaking up water. These cells enable the moss to grow in open or standing water. During periods of drought, mosses can release excess water and still survive.

Sphagnum mosses are significant to wetlands because they help create a unique environment.

Sphagnum mosses create bogs. They can grow out and over pools of standing water. Over long periods of time, they completely cover the pool and form a floating mat – growing on top and dying below. The bog mat can grow to hundreds of feet in depth. Larger, open bogs are found further north in the Great Lakes region and around the world in northern latitudes.

Mosses can store water - up to twenty times their dry weight. Because they can take up so much water, the surrounding plants are starved of the nutrients they need. As a result, only a few plant species can survive with this moss. Most mosses grow only a few inches high. The stunted shrubs and trees in bogs have adapted to these poor growing conditions and are able to survive.

Mosses form organic soil when slowly decaying moss far below the surface decomposes. The result is highly acidic and nutrient-poor soil.

You may encounter a small relict of a bog in the mosaic of the cedar swamp forest. During your wetland walk, look for these common bog plants in association with sphagnum mosses:

Shrubs: leatherleaf, Labrador tea, bog rosemary, bog-cranberry

Flowers: northern pitcher plant, round-leaved sundew, rose pogonia, three-leaved Solomon's seal

Feel the soft branches of the moss you find. If you gently squeeze a handful of moss, it will release lots of water which is critical to wetland plant survival. Peat moss is harvested, dried and sold in nurseries for use in gardening. When peat moss is added to sandy soil, it holds moisture and helps to prevent plant roots from drying out.

Sphagnum moss (*Sphagnum andersonianum*)

Sphagnum moss Family: Sphagnum Moss (Sphagnaceae)

Scientific Name: Sphagnum section Sphagnum Height: 4" (10cm)

SPORES: capsule about 1 cm above ground

LEAF: Evergreen, hood-shaped, most species light green, tiny, slightly toothed, scattered along stems with living and dead cells

STEM: Evergreen, tightly arranged clusters of 2-3 spreading branches and 2-4 hanging branches, form large hummocks

NOTE: Plants can hold 16-26 times as much water as their weight. Native Americans used this moss to dress wounds. Sphagnum moss species are grouped into 4 sections. Sphagnum sect. Sphagnum most common in Upper Great Lakes.

FERN ALLIES CHECKLIST

- O Dwarf scouring rush
- O Northern fir-moss
- O Shining fir-moss
- O Sphagnum moss
- O Tree clubmoss
- O Water horsetail

FIELD NOTES

RESOURCES

PROTECTING WETLANDS

Human activities have destroyed **more than 90%** of our vital wetlands,

BUT you can help protect our remaining wetland communities.

BE PROACTIVE!

Action #1- Clean your shoes before walking in wetlands.

The crevices in the soles of your shoes can transport seeds from wetlands to your property. By taking the time to wash off the bottoms of your shoes, you can prevent the spread of invasive species.

Action #2- Remove invasive species from your land.

For example, invasive purple loosestrife grows in wet places: along your driveway, beside the brook meandering through your property, among the cattails across the road from your home.

It may be pretty, but a single plant produces more than one million seeds each year. In a short time, these plants will choke out all of our native plants.

Removing a single plant prevents the spread of this non-native species and helps preserve natural habitat for our native birds, bees and butterflies!

Action #3- Maintain wetlands on your property.

Maintain a wetland instead of filling, dredging or draining it.

Wetlands keep our water clean by filtering stormwater runoff, so incorporate them into your landscaping.

Install more native wetland plants, because they improve habitat for native species like fireflies and dragonflies.

Action #4- Clean, Drain and Dry your Watercraft.

Enjoy your boat or kayak observing wetlands along a river or along the lakeshore. Then take time to clean, drain & dry your watercraft, trailer and gear.

For example, invasive Eurasian milfoil clings to boat trailers and propellers. Their microscopic seeds attach to kayak paddles and bottoms of your watershoes. Milfoil grows so rapidly that shallow lakes can become impassable in just a few years.

Action #5- Learn how to identify and remove invasive species.

You can contact your state or provincial Department of Environment or local Conservation District office.

<u>United States- Great Lakes Basin</u>

https://www.epa.gov/greatlakes/invasive-species-great-lakes

https://www.regions.noaa.gov/great-lakes/index.php/great_lakes-restoration-initiative/

https://habitatmatters.org (NW Michigan Invasive Species Network)

<u>Canada</u>

https://www.invasivespeciescentre.ca

<u>Show you care!</u>

FIELD NOTES

THE AMAZING WORLD OF FLOWERS!

How did flowers develop?

Loren Eiseley, a famous anthropologist, wrote an essay in 1946 called, "How flowers changed the world." He started by writing, "Once upon a time, there were no flowers at all." Can you imagine a world without flowers?

"Flowers changed the face of our planet. Without them, the world we know - even man himself - would never have existed... Flowering plants emerged before the close of the Age of Reptiles. The agile brains of the warm-blooded birds and mammals demand a high oxygen consumption and food in concentrated forms... It was the rise of the flowering plants that provided that energy and changed the nature of the living world. Their appearance parallels the rise of the birds and mammals."

"Our present day pine forests represent plants of a pollen-disseminating variety. Once fertilization was no longer dependent on water, the march over drier regions could be extended. The trees are ancient, slow-growing and immense, like the redwood groves that have survived to our day on the California coast...There was no grass as yet, no wide plains rolling in the sun, no tiny daisies dotting the meadows. There is little versatility..."

"...the true flowering plants (angiosperm itself means "encased seed") grew a seed in the heart of a flower, a seed whose development was initiated by a fertilizing pollen grain independent of outside moisture. But the see, unlike the developing spore, is already a fully equipped embryonic plant packed in a little enclosed box stuffed full of nutritious food... by feather down attachments, as in dandelion or milkweed seed, it can be wafted upward on gusts and ride the wind for miles; or with hooks it can cling to a bear's or rabbit's hide; or like some of the berries, it can be covered with a juicy, attractive fruit to lure birds, pass undigested through their intestinal tracts and be voided miles away... Plants traveled as they had never traveled before."

"Food came from three sources, all produced by the reproductive system of the flowering plants:

- Tantalizing nectars and pollen intended to draw insects for pollinizing purposes and hummingbirds
- Juicy and enticing fruits to attract larger animals, and in which tough-coated seeds were concealed, as in the tomato
- Food in the actual seed itself"

"Grass began to cover the bare earth, until today, there are over 6,000 species." And the rest is history…

PLANT LIST

Common Name	Scientific Name
Alder, Tag/Speckled	*Alnus rugosa*
Arbutus, Trailing	*Epigaea repens*
Arrowhead, Northern	*Sagittaria latifolia*
Ash, Black	*Fraxinus nigra*
Aster, Purple-stem	*Symphytotrichum punicem*
Avens, Purple	*Geum rivale*
Bedstraw, Rough	*Galium asprellum*
Bellflower, Marsh	*Campanula aparinoides*
Birch, Yellow	*Betula alleghanensis*
Bladderwort, Common	*Utricularia vulgaris*
Boneset	*Eupatorium perfoliatum*
Bulrush, Softstem	*Schoenoplectus tabernaemontani*
Bunchberry	*Cornus canadensis*
Bur-reed, Common	*Sparganium americanum*
Buttercup, Northern swamp	*Ranunculus hispidus*
Buttonbush	*Cephalanthus occidentalis*
Canary-grass, Reed	*Phalaris arundinacea*
Cardinal flower	*Lobelia cardinalis*
Cattail, Broad-leaved	*Typha latifolia*
Cattail, Narrow-leaved	*Typha angustifolia*
Cedar, Northern White	*Thuja occidentalis*
Cinquefoil, Marsh/Purple	*Comarum palustris*
Cinquefoil, Shrubby	*Dasiphora fruticosa*
Clintonia	*Clintonia borealis*
Clubmoss, Tree	*Dendrolycopodium spp.*
Cranberry, Small	*Vaccinium oxycoccus*
Dogwood, Red-twig	*Cornus sericea*
Duckweed, Common	*Lemna minor*
Elderberry, Common	*Sambucus canadensis*
Fern, Bulblet	*Cystopteris bulberifera*

Fern, Cinnamon	*Osmunda cinnamomea*
Fern, Interrupted	*Osmunda claytoniana*
Fern, Lady	*Athyrium filix- femina*
Fern, Marsh	*Thelypteris palustris*
Fern, Oak	*Gymnocarpium dryopteris*
Fern, Ostrich	*Matteuccia struthiopteris*
Fern, Royal	*Osmunda regalis*
Fern, Sensitive	*Onoclea sensibilis*
Fir, Balsam	*Abies balsamea*
Fir-moss, Northern	*Huperiza selago*
Fir-moss, Shining	*Huperiza lucidula*
Goldenrod, Bog	*Solidago uliginosa*
Goldenrod, Rough-leaved	*Solidago patula*
Goldenrod, Smooth	*Solidago gigantea*
Goldthread, Three-leaf	*Coptis trifolia*
Grass of Parnassus	*Parnassia glauca*
Grass Pink	*Calopogon tuberosus*
Horsetail, Water	*Equisetum fluviatile*
Iris, Blue Flag	*Iris versicolor*
Iris, Yellow Flag	*Iris pseudacorus*
Jack-in-the-Pulpit	*Arisaema triphyllum*
Jewelweed	*Impatiens capensis*
Joe Pye-Weed, Spotted	*Eupatorium maculatum*
Labrador Tea	*Ledum groenlandicum*
Ladies tresses, Nodding	*Spiranthes cernua*
Lady's Slipper, Large Yellow	*Cypridium calceolus var. pubescens*
Lady's Slipper, Small Yellow	*Cypridium calceolus var. parviflorum*
Lady's Slipper, Pink	*Cypridium acaule*
Lady's Slipper, Showy	*Cypridium reginae*
Leatherleaf	*Chamaedaphne calyculata*
Lily, Wood	*Lilium philadelphicum*
Lobelia, Brook	*Lobelia kalmii*
Lobelia, Great Blue	*Lobelia siphilitica*
Loosestrife, Purple	*Lythrum salicaria*
Loosestrife, Smooth	*Lysimachia quadriflora*
Loosestrife, Tufted	*Lysimachia thyrsiflora*

Loosestrife, Yellow	*Lysimachia terrestris*
Maple, Red	*Acer rubrum*
Marigold, Marsh	*Caltha palustris*
Mayflower, Canada	*Maianthemum canadense*
Meadow-Rue, Purple	*Thalictrum dasycarpum*
Meadowsweet	*Spirea alba*
Milkweed, Swamp	*Asclepias incarnata*
Mitrewort, Nude	*Mitella nuda*
Moss, Sphagnum	*Sphagnum spp.*
Nettle, Stinging	*Utica dioica*
Nightshade, Deadly	*Atropa belladonna*
Orchid, Calypso	*Calypso bulbosa*
Orchid, Dragon's Mouth	*Arethusa bulbosa*
Orchid, Northern green bog	*Platanthera aquilonis*
Orchid, Small purple fringed	*Platanthera psycodes*
Orchid, Tall white bog	*Platanthera dilatata*
Pickerel Weed	*Pontederia cordata*
Pitcher Plant, Northern	*Sarracenia purpurea*
Pogonia, Rose	*Pogonia ophioglossoides*
Poison Ivy	*Toxicodendron radicans*
Poison Sumac	*Toxicodendron vernix*
Polygala, Fringed/Gaywings	*Polygala paucifolia*
Ragwort, Golden	*Packera aurea*
Raspberry, Dwarf	*Rubus acaulis*
Reed, Common Non-Native	*Phragmites australis v. australis*
Reed, Common Native	*Phragmites australis v. americanus*
Rose, Swamp	*Rosa palustris*
Rosemary, Bog	*Andromeda glaucophylla*
Scouring-Rush, Dwarf	*Equisetum scirpoides*
Sedge, Porcupine	*Carex hystericina*
Sedge, Tussock	*Carex stricta*
Snowberry, Creeping	*Gaultheria hispidula*
Spruce, Black	*Picea mariana*
Spurge, Leafy	*Euphorbia esula*
Starflower	*Trientalis borealis*
Sundew, Round-leaved	*Drosera rotundifolia*

Sweet Gale	*Myrica gale*
Tamarack	*Larix laricina*
Thistle, Marsh/European	*Cirsium palustre*
Thistle, Swamp	*Cirsium muticum*
Turtlehead	*Chelone glabra*
Twinflower	*Linnaea borealis*
Vervain, Blue	*Verbena hastata*
Water- Hemlock, Bulb-bearing	*Cicuta maculata*
Water- Hyacinth	*Eichornia crassipes*
Water -Lily, Bullhead	*Nuphar variegata*
Water- Lily, White	*Numphaea odorata*
Water- Milfoil, Common	*Myriophyllum sibiricum*
Water- Milfoil, Eurasian	*Myriophyllum spicatum*
Water- Smartweed	*Persicaria amphibia*
Water- Weed, Common	*Elodea canadensis*
Willow, Pussy	*Salix discolor*
Winterberry	*Ilex verticillata*
Wintergreen, Common	*Gaultheria procumbens*

GLOSSARY

Alternate leaves – a leaf arrangement where leaves grow at different levels along the stem of a plant

Anther – the part of a flower that carries pollen

Bilateral symmetry – a flower can be divided into roughly mirror–image halves

Bract – a structure at the base of a flower that looks like a small leaf

Bud – an undeveloped or unopened flower

Clasping leaf – a leaf that encircles the stem of a plant

Composite flower – a flower with a head that brings together several to very many tiny flowers, so they look like one large flower

Compound leaf – a leaf with two or more leaflets

Corolla – all of the parts of a flower

Evergreen – a plant that keeps its leaves year round

Fiddlehead – the furled fronds of a young fern

Filament – the stalk that holds the male reproductive part of a plant (the stamen) and supports the pollen–bearing part of a plant (the anther)

Frond – a fern leaf

Involucre – a whorl of bracts below a flower

Irregular flower – a flower without radial symmetry

Leaf axil – the point of the angle between a leaf and a stem

Leaf blade – the flat part of a leaf

Leaflet – one of the leaf–like structures in a compound leaf

Leaf margin – the outer edge of a leaf

Opposite leaves – a leaf arrangement in which leaves are paired opposite one another on a stem

Ovary – the lower part of the female reproductive part of a flower (the pistil) that produces the seeds

Palmate leaf – a leaf that is divided like the fingers of a hand

Perfoliate – a leaf that has the stem apparently passing through the leaf

Petal – an individual part of the flower that is often white or colored

Pinnate leaf – a leaf that is divided once along a central axis into separate segments

Pistil – the female part of the flower that produces seeds and consists of an ovary and one or more styles (stalks) and stigmas

Pollen – the male spores located in the male reproductive part of a flower (the anther)

Radial symmetry – roughly identical flower parts occuring at regular intervals around the axis of the flower

Regular flower – a flower with all of the members of a single whorl, such as petals, are similar in shape and size

Rhizome – an underground stem of a plant that grows horizontally

Sepals – one of the separate parts of the flower that encloses the petals, usually green

Simple leaf – a leaf that is not divided

Spore – one cell on a fern that is capable of producing a new individual plant

Stalk (flower/leaf) – the part of a flower or leaf that attaches to the stem of a plant

Stamen – the male part of a flower that produces pollen and consists of a stalk bearing an anther

Stem – the main upright part of a plant

Stigma – the top of the female reproductive part of a flower that receives pollen

Stipule – a leaf–like structure at the base of a leaf stalk

Style – the stalk–like part of a female reproductive part of a flower

Whorl – an arrangement of leaves that grow in rings around the stem of a plant

Winged stem – a thin, leaf–like structure growing along a stem

INDEX

BOLD NUMBER - Page with photo

A

B

C

N

O

P

R

Y

FIELD NOTES

FIELD NOTES

FIELD NOTES

FIELD NOTES

FIELD NOTES

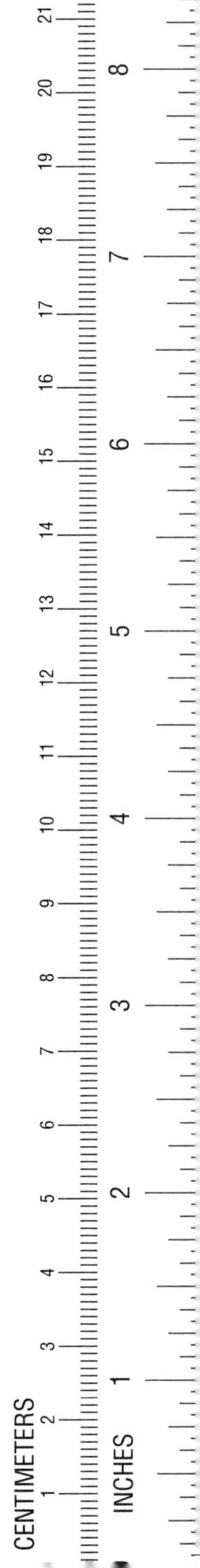
CENTIMETERS
INCHES

Peg Comfort has a passion for water, wilderness and loons. She holds a BA in American Studies from Purdue University and an MS in Environmental Studies from Bemidji State University. As an educator, naturalist and writer, Peg has dedicated her life to conservation. After serving as director of the ISEA Global Freshwater Initiative, she began to think globally and act locally. She volunteered on the first board of directors for the Grass River Natural Area (GRNA), Inland Seas Education Association (ISEA) and Torch Conservation Center (TCC). Peg and her husband built their home and raised their son on Torch Lake.

Grass River Natural Area Inc. is a non-profit organization in Bellaire, Michigan with a mission to "manage the Grass River Natural Area, conserve and protect its watershed and provide opportunities that increase knowledge, appreciation, and community-wide stewardship of the natural environment." See more at www.grassriver.org

James Dake is an educator, naturalist, and photographer originally from Flushing, Michigan. He is the education director at Grass River Natural Area and currently lives in Elk Rapids, Michigan. He is the author of *Field Guide to Northwest Michigan*, a GRNA publication.

Photo credits: Except for the images noted below, all photographs are by James Dake.

> Aerial Wetland Photo p. 225 (Andy Zendler); Panoramic Habitat Photos (MNFI/GRNA); Northern Fir-moss [Erlend Bjørtvedt (CC-BY-SA 3.0)]; Water Hyacinth [Ravi Jung Karki (CC-BY-SA 4.0)]; Great Lakes Basin Map [drdpw (CC-BY-SA 3.0)]

GRNA would like to offer a special thank you to those who made this publication possible: Peg Comfort, Sue Schuler, Grand Traverse Regional Community Foundation, Glenn Wolff, Jerry Dennis, Heather Smith, Jenn Wright, Sheila Fox, Betsy Willis, Arlene Westhoven, Emily Burke, Liam Fries, Valerie Reeves, Anne Drake, Franny Bluhm, Brian Thelan, and all of the passionate individuals who have contributed to the protection of the Grass River Natural Area during its first 50 years.